C. Cattaneo (Ed.)

Relatività generale

Lectures given at the
Centro Internazionale Matematico Estivo (C.I.M.E.),
held in Salice d´Ulzio (Torino), Italy,
July 16-25, 1964

C.I.M.E. Foundation
c/o Dipartimento di Matematica "U. Dini"
Viale Morgagni n. 67/a
50134 Firenze
Italy
cime@math.unifi.it

ISBN 978-3-642-11020-7 e-ISBN: 978-3-642-11021-4
DOI:10.1007/978-3-642-11021-4
Springer Heidelberg Dordrecht London New York

Reprint of the 1st ed. C.I.M.E., Ed. Cremonese, Roma, 1965
With kind permission of C.I.M.E.

Printed on acid-free paper

Springer.com

CENTRO INTERNATIONALE MATEMATICO ESTIVO
(C.I.M.E)

Reprint of the 1st ed.- Salice d' Ulzio, Italy, July 16-25, 1964

RELATIVITÀ GENERALE

PREFACE

The following lectures were intended to serve as an introduction to the theory of gravitational waves, mainly for mathematicians not specialized in the field of general relativity. Accordingly, basic concepts and motivations an the purely local, differential geometrical "pure" radiation theory have been put in the foreground, and conceptually and computationally more complicated recent advances have indicated only briefly.

The references and footnotes should be considered an essential part of the course; I hope that some of them serve to clarify points raised in discussions which followed the lectures.

GRAVITATIONAL WAVES

by Jürgen Ehlers

1. Introduction : The Basis of the General Theory of Relativity

From a physicist's point of view the general theory of relativity is of basic importance, despite of its very poor experimental or observational verification, for two reasons :

a) It is the most convincing field theory of gravitation which is locally compatible with the experimentally well-established Lorentzian structure of the space-time metric, and

b) it is the most important example of a physical theory in which the metric structure of space-time is treated not as given a priori, but dependent on and interrelated to other physical variables describing processes in space-time.

Although b) is not independent of a) it is worthwhile to stress the autonomous importance of aspect b): So far, every physical theory, whether non-relativistic or relativistic, classical or quantum, whether a particle -or a field theory, requires for the formulation of its basic laws as well as for its interpretation a metric and, associated with it, an affine connection which serves to formulate laws relating quantities with directional properties at different space-time points or "events". This implies that in all physical theories the metric has a strong influence on other physical quantities - I need only mention the law of inertia so fundamental not only for classical mechanics but also for, say, the quantum theory of scattering. Nevertheless this metric structure is not reinfluenced by these physical quantities except in the general theory of relativity and its generalizations. This strongly suggests the idea that the pre-Einsteinian theories may well be considered as approximate theories which describe situations in which the metric field can be treated as an external field which has, under the special circumstances considered, always the same structure, whereas in more general cases or in a more precise description the metric is a field variable like, say, the electromagnetic field. It is certainly more convincing to have a theory where all quantities which are used to interpret the observed phenomena are interrelated ("principle of omnipresence of all state variables", to use a phrase from the modern theory of irreversible processes in continuous media) than the assign some of these quantities a priori and prescribe "laws" only for the remaining ones.

If this point of view is accepted, then the gravitational field - if it is identified with the metric field - acquires, despite of its extreme weakness even in comparison with so called "weak" interactions, a fundamental role in physics since it is coupled to all other fields, due to the role of the metric stressed above. There is a very good reason for this identication, namely the universal proportionality of inertial and ("passive") gravitational mass of bodies substantiated with a precision of 10^{-11} by the Eötvös-Dicke experiment[1].

Let us, then, accept this idea of the metric as a physical field, and formulate the first basic assumption of the Einsteinian theory, motivated by the special theory of relativity :

(G) Riemannian assumption : The space-time manifold $V_4 = V$ carries a normal-hyperbolic Riemannian metric with the fundamental quadratic form (in an arbitrary local coordinate-system)

$$G = g_{ab}(x^c)dx^a dx^b \quad . \qquad (1 \leqslant a, b, \ldots \leqslant 4)$$

(We take the signature to be +++-.)

Since g_{ab} is supposed to describe the gravitational field well-known considerations of Einstein (which contain some weak points which are still not clarified completely) lead to the assumption that the source of the g_{ab} -field is the stress-energy-momentum tensor of all the matter populating V ; we formulate the second assumption :

(T) The mechanical properties of matter are described by a symmetric tensor field T_{ab} depending on the state-variables of matter.

Finally, Newtonian theory (Poisson's equation), simplicity-requirements (quasi-linearity, second differentiation order for the g_{ab} 's) and energy-momentum conservation[2] suggest the most specific assumption of Einstein's 1915-theory :

(GT) The metric G of space time is related to the stress energy momentum distribution T in V by the field equation[3]

(1) $$G_{ab} + T_{ab} = 0 \quad .$$

Here the Einstein tensor $G_{ab} = R_{ab} - \frac{1}{2} g_{ab} R$ occurs where R_{ab} is the contracted curvature tensor, and R its trace. (We choose units such that $c = 1$ and (Newton's constant of gravity) $= \frac{1}{8\pi}$.)

The gravitational field equation (1) is, of course, not sufficient as a basis for a theory of the interaction between matter and the gravitational field. It is necessary to add assumptions about the structure of matter, i.e. to specify the dependence of T_{ab} on the basic matter (or field) variables, and to state the non-gravitational equations of motion which these variables are supposed to obey. Since, however, the interaction between matter and gravitational waves has so far not been investigated in the full, non-linear theory, we need not specify such assumptions here.

In empty space, where (1) reduces to

(2) $$G_{ab} = 0, \qquad \text{or} \qquad R_{ab} = 0 \quad ,$$

no further assumptions are needed in order to calculate the time-development of a field from given initial data.

The gravitational field equation implies the "mechanical law"

$$T^{ab}{}_{;b} = 0 \tag{3}$$

which is the general-relativistic analogue of the balance-equations for energy and momentum for continuous media.

For isolated bodies of appropriate internal structure one can "approximately deduce" from (3) equations of motion for the center of mass world line and for multipole moments describing the structure of the body such as spin, quadruple moment etc. We adopt here[4] as equations of motion of an approximately rigid, spherically symmetrical test particle with internal angular momentum per proper mass S^a :

$$(4)_1 \qquad \frac{dx^a}{ds} = u^a \, , \quad \frac{\nabla u^a}{ds} + R^{*a}{}_{bcd} u^b u^d S^c = 0 \quad ,$$

$$(4)_2 \qquad \frac{\nabla S^a}{ds} (\delta^b_a + u_a u^b) = 0 \quad .$$

s denotes the proper time, u^a the 4-velocity, $u_a u^a = -1$, $\frac{\nabla u^a}{ds}$ is the 4-acceleration, and

$$R^{*\,abcd} = \frac{1}{2} R^{ab}{}_{ef} \eta^{efcd}$$

is the "right-dual" of the Riemann curvature tensor. The metric quantities are those of the external field.

A nonspinning test particle has, according to $(4)_1$, a geodesic world line. For a pair of neighbouring nonspinning test particles the relative acceleration is a linear transform of the relative position vector x^a, $u_a \delta x^a = 0$:

$$\frac{\nabla^2 \delta x^a}{ds^2} = R^a{}_{bcd} u^b u^d \delta x^c \tag{5}$$

These equations of motion for test particles give direct operational meaning to the metric g_{ab} since the set of all timelike geodesics determines a normal hyperbolic metric uniquely up to a constant factor[5]. Moreover, $(3)_1$ and (4) give a precise meaning to the statement that "the curvature tensor describes the strength and the directional properties of a gravitational field similarly to the way in which the field strength tensor describes an electromagnetic field.

Finally, we observe that $(4)_2$ gives a physical meaning to the Fermi propagation of vectors along curves. Since we may take the spin as small as we like[6] for a given mass, we can, to any desired degree of accuracy, realise a geodesic with a vector parallely propagated along it and orthogonal to the curve. Taking two such test-gyroscopes near one another, we can supplement (5) by the statement[7]:

The difference δS^a between tha angular momentum of the first particle and that of the second particle parallel displaced along the connection vector δx^a and projected into the local space orthogonal to the 4-velocity u^a of the first particle, $\delta_1 S^a$, obeys the law

$$\frac{\nabla \overrightarrow{\delta_1 S^a}}{ds} = \vec{S} \times \vec{H}, \tag{6$_1$}$$

where

$$H^a = R^{*a}{}_{bcd} u^b u^d \delta x^c . \tag{6$_2$}$$

Here $\delta_{\perp} S^{a}$, S^{a} , H^{a} belong to the 3-space orthogonal to u^{a} , and $(6)_1$ is written as a 3-vector relation containing the usual exterior product. Rewritten in this notation, (5) assumes the form

$$(5)_1 \qquad \frac{\nabla^2 \overrightarrow{\delta x}}{ds^2} = \vec{E} ,$$

$$(5)_2 \qquad E^{a} = R^{a}{}_{bcd} u^{b} u^{d} \delta x^{a} .$$

The equations (6) and (5') exhibit that, for a given "observer" u^{a} , the (spatial) vectorfields $\vec{E}$ and $\vec{H}$ defined in the infinitesimal neighbouhood of the observer's world line play a similar role for a gravitational field as the electric and magnetic vectors relative to an inertial frame for an electromagnetic field.

A null-geodesic also has a physical interpretation : It represents the world line of a particle of vanishing rest mass or, more classically, a light ray in the sense of geometrical optics. In this case, the statement can be "approximately deduced" by starting with the general relativistic form of Maxwell's equations and going over to the limit of "locally plane waves of infinitely small wave-length"[8].

As long as we do not have a description of the interaction of matter with gravitational fields, especially gravitational waves, the preceding remarks on test-body motions are a useful preliminary tool for the physical interpretation of algebraic and anlytic properties of vacuum gravitational fields and, especially, their curvature tensors. One should keep in mind, however, that this description of the action of gravitational fields on matter is very incomplete since the reaction of the particles on the fields is completely neglected.

2. The linear approximation. Survey of problems

In order to get a survey over the problems with which we are faced let us at first drastically eliminate the mathematical complications due to the non-linearity of eqs. (1) :

Let us denote by Δ_{ab} the orthonormal components of the flat space time metric, and let us assume that the quantities

(7) $$\eta_{ab} \equiv g_{ab} - \Delta_{ab}$$

satisfy the "weak field conditions"

(8)$_1$ $$\left|\eta_{ab}\right| << 1 \quad ,$$

(8)$_2$ $$\left|\Gamma^{e}_{b[c} \Gamma^{a}_{d]e}\right| << \left|\Gamma^{a}_{b[c,d]}\right| \quad ,$$

where Γ^{a}_{bc} are the Christoffel symbols associated with the g_{ab} . Then the field equation (1) reduces, in the sense of a formal approximation in which small quantities are neglected, to the linearised field equation

(9) $$\Box \psi_{ab} - 2\psi_{(a,b)} + \Delta_{ab} \psi^{c}{}_{,c} = -2T_{ab}$$

here

(10) $$\psi_{ab} \equiv \eta_{ab} - \frac{1}{2} \Delta_{ab}\eta \quad , \quad \eta \equiv \eta^{a}{}_{a}, \; \psi_{a} \equiv \psi_{a}{}^{b}{}_{,b},$$

and the D'Alembert-operator $\Box$ and the raising and lowering of indices refer to the flat metric Δ_{ab} .

A "small" coordinate change

(11) $$x^{a'} = x^a + \xi^a \quad , \quad \left|\xi^a{}_{,b}\, \xi^c{}_{,d}\right| \ll \left|\xi^e{}_{,f}\right|$$

induces the transformation

$(12)_1$ $$\psi_{a'b'} = \psi_{ab} - 2\,\xi_{(a,b)} + \Delta_{ab}\,\xi^c{}_{,c} \quad ,$$

$(12)_2$ $$\psi_{a'} = \psi_a - \Box\,\xi_a$$

of the field variables ψ_{ab} .

One may now forget the "derivation" of (9) and (12) from the rigorous theory and consider (9) as a gravitational field equation in flat space-time, formally very similar to electrodynamics. Then (12) can be considered not as induced by a coordinate transformation bur as a gage-transformation; in fact, the substitution (12) (with unchanged independent variables x^a !) leaves the left hand side of eq. (9) unchanged. It also follows from (9) that

(13) $$T^{ab}{}_{,b} = 0 \quad .$$

But this equation clearly shows this linear theory of gravitation being physically wrong : According to (13), the gravitational field ψ_{ab} would have no influence on the energy and momentum balances of matter. Although the field is determined by its source T_{ab} only up to gage transformations the linearised equation of motion of a test particle is not gage invariant; this is a second inconsistency[9].

We therefore have to consider (9) at best as the first step in a sequence of successive approximations[10]. Let us nevertheless apply the flat-space interpretation of (9), (12) in the following and state some mathematical properties of this theory rigorously, as a motivation for the analysis of the full theory.

Let us consider a spatially bounded source T^{ab} at rest in some inertial frame. Then the retarded integral

$$(14) \qquad \psi_{ab}(x) = \frac{1}{2\pi} \int_{C_x^-} T_{ab} \, dK$$

exists and satisfies, if (13) holds, the Einstein convention

$$(15) \qquad \psi_a = 0$$

and the field equation (9). (dK is the Lorentz-invariant measure on the past light cone C_x^- of x.) If T^{ab} and its first derivatives are bounded in the past, (14) satisfies the boundary conditions

$$\psi_{ab} = \theta\left(\frac{1}{r}\right), \qquad \psi_{ab,c} = \chi_{ab} k_c + \theta\left(\frac{1}{r}\right),$$

$$\chi_{ab} = \theta\left(\frac{1}{r}\right), \qquad \psi_a = \theta\left(\frac{1}{r^{z+\varepsilon}}\right), \qquad \varepsilon > 0 .$$

r denotes the spatial distance of the argument of $\psi_{..}$ from a time like straight line contained in the source region, and k^a is a null vector field pointing away from the source and into the future, normalised according to

$k^a u_a = -1$ if u^a is the 4-velocity of the line mentioned above.

We now state the

theorem [11] : For a given source T^{ab} exists up to gage transformations one and only one solution ψ_{ab} of (9) which satisfies the "outgoing radiation condition" (16). Among these, precisely one satisfies the Einstein convention (15) .

To prove uniqueness, we apply the Kirchhoff integral representation [12] (known in physics from the theory of diffraction)

(17) $$4\pi \cdot \psi_{ab}(x) = -\int_{C_x^-} \Box \psi_{ab} \, dK$$

to the difference of two solutions of (9) both satisfying (16). The surface integrals in the general Kirchhoff-representation can and have been shifted to (past) infinity in C_x^- and then give zero because of $(16)_{1,2,3}$. From (9) and (17) we obtain for this difference $4\pi \, \psi_{ab}(x) = -\int_{C_x^-} (2\, \psi_{(a,b)} - \Delta_{ab} \psi^{c}_{\ ,c}) d$ which can be written, on account of $(16)_4$, in the form $-2\, \xi_{(a,b)} + \Delta_{ab}\, \xi^c_{\ ,c}$ with $\xi_a(x) = \int_{C_x^-} \psi_a \, dK$, and is, consequently, gage-equivalent to zero. The existence has already been shown.

A motivation for the name "outgoing radiation condition" for (16) can be seen in the fact that the change

$$d\,\psi_{ab} = \chi_{ab} k_c dx^c + \theta\left(\frac{1}{r^2}\right)_c dx^c$$

at large distances from the source is smallest for displacements within the hypersurfaces of constant phase, $k_a dx^a = 0$.

Because of this theorem, it is no loss of generality for problems involving bounded sources only to impose generally the condition (15), i.e.

$$\psi_{a\,;b}^{\;b} = 0 \,. \tag{18}$$

Then (9) simplifies to

$$\Box\psi_{ab} = -2T_{ab} \tag{19}$$

Outside of the sources, we have

$$\Box\psi_{ab} = 0, \qquad \psi_{a\,,b}^{\;b} = 0 \,, \tag{20}$$

and we define, in the linear approximation, "free" gravitational waves as gage-equivalente classes of solutions of (20).

The "free" classical field theory defined by (20) can be used to construct a corresponding special relativistic quantum theory of a "graviton field". For this purpose one hase to define, on a suitably chosen subset of the solutions of (20), a Hilbert space structure with a scalar product that is invariant under (inhomogeneous) Lorentz transformations. You obtain thus an irreducible unitary representation of the inhomogeneous Lorentz group in a Hilbert space of solutions of (20) which is to be interpreted physically as the space of one - graviton states. According to the group theoretic classification of fundamental particles (or fields), one then finds the linearised free graviton field belonging to particles with vanishing rest mass and spin 2. (The spaces of n-particle states and, finally, the total (Fock-) space of the free graviton field can be constructed by standard procedures from the space of one-particle states.)

The metric corresponding to the general solution of (20) which represents a plane wave travelling in the z-direction can be written in the form

$$G = \underset{0}{G} + A(dx^2 - dy^2) + 2B\,dxdy \tag{21}$$

with two arbitrary functions A, B of the "phase" $u = z-t$. $\underset{0}{G}$ denotes the Minkowskian metric.

Since the coefficients of (21) depend on u only, the curvature tensor (and all intrinsic characteristics of the metric field) are propagated without change along the rays (x, y, u) = const. which form a congruence of null geodesics, i.e. a plane gravitational wave propagates without distortion with fundamental velocity.

(21) is in Gauss'normal form with respect to t, and thus the geodesics (x, y, z) = const. may bethought of as world lines of test particles. It follows from (21) that a cloud of such particles undergoes a volume-preserving deformation which is restricted to directions orthogonal to the direction of propagation of the wave; the magnitude of this deformation depends on the amplitudes A, B.

In order to characterize the wave (21)independently of a special set of test particles we use the linearized curvature tensor. It has the form

$$R_{abcd} = m_{ab}m_{cd} - \overset{*}{m}_{ab}\overset{*}{m}_{cd} \tag*{$(22)_1$}$$

where m_{ab} is a singular bivector,

$$m_{ab}m^{ab} = 0 \quad , \quad \overset{*}{m}_{ab}m^{ab} = 0 \quad . \tag*{$(22)_2$}$$

Consequently there exists a null vector k^a such that

$$(22)_3 \qquad m_{ac} m_b{}^c = k_a k_b \qquad (\Longrightarrow m_{ab} k^b = 0).$$

The quantities m_{ab}, k^a are determined by R_{abcd} up to their signs.

The interpretation of m_{ab}, k^a follows from eq. (5) : Let u^a be the 4-velocity of an arbitrary nonspinning test particle or "observer" . Then

$$(23)_1 \qquad p^a \equiv \frac{m^a{}_b u^b}{-k_c u^c} \quad , \qquad q^a \equiv \frac{\overset{*}{m}{}^a{}_b u^b}{-k_c u^c}$$

form an orthogonal pair of (with respect to this observer) purely spatial vectors, and (22) may be rewritten as

$$(23)_2 \qquad R_{abcd} = 4\,k_{[b} i_{a][c} k_{d]} \; ,$$

$$(23)_3 \qquad i_{ab} \equiv p_a p_b - q_a q_b \; .$$

The accelerations of nearby test particles relative to our observer are, according to (5) and (23) , given by

$$(24) \qquad \frac{\nabla^2 \delta x^a}{ds^2} = (k_c u^c)^2 \, i^a{}_b \, \delta x^b \; .$$

These formulae show : The acceleration of a test particle relative to a freely falling observer vanishes if and only if its position vector δx^a is parallel to the projection $k_\perp^a$ of k^a into the observers 3-space. $(k_a u^a)^2$ is equal to the ratio (magnitude of relative acceleration / distance) for arbitrary nearby test particles. The acceleration is parallel to δx^a if $\delta x^a = \lambda p^a$, anti-parallel if $\delta x^a = \lambda q^a$.

For the wave (21) k^a is given by

$$k_a dx^a = \left(\frac{1}{4}(A''^2 + B''^2)\right)^{1/4} du \qquad (25)$$

and the propagation of the wave along the rays is expressed by

$$m_{ab;c} k^c = 0 \qquad (26)_1$$

which implies, by (22),

$$R_{abcd;e} k^e = 0, \qquad k_{a;b} k^b = 0 \quad . \qquad (26)_2$$

A freely falling observer, however, will notice changes of the field; the strength $(k_a u^a)^2$ will be a function of this proper time, and the directions $k_\perp^a$, p^a, q^a will rotate relative to spatial axes which are parallely propagated along his world line. These changes may be used to define, with respect to an observer, monochromatic waves and, among them, linearly, circularly etc. polarized waves quite similar to electrodynamics.

We finally remark that vacuum curvature tensors of the algebraic type (22) can be characterized by the existence of a vector k^a such that

$$R_{abcd} k^d = 0 \qquad (\Rightarrow \; k_a k^a = 0) \; ; \qquad (27)$$

this remark suggests a way of defining pure radiation fields in the rigorous theory.

Let us now return to the inhomogeneous equation (19) and its retarded solution (14). If we choose an inertial frame in which the source is at rest and located near the origin of the space-coordinates we may write

$$(28)\qquad \psi_{ab}(\underline{x},t) = \frac{1}{2\pi}\int \frac{T_{ab}(\underline{y}, t - |\underline{x}-\underline{y}|)}{|\underline{x}-\underline{y}|}\, d^3y$$

($\underline{x}$, $\underline{y}$ denote 3-vectors). If we are interested in the radiation field at large distances from the source, we will, as usual, write $t - |\underline{x}-\underline{y}| = t - |\underline{x}| + (|\underline{x}| - |\underline{x}-\underline{y}|)$ and develop $\frac{1}{|\underline{x}-\underline{y}|}$ and $|\underline{x}| - |\underline{x}-\underline{y}|$ in powers of r^{-1}, obtaining

$$(29)_1\qquad \psi_{ab}(\underline{x},t) = \frac{N_{ab}(u,\omega)}{r} + \theta\left(\frac{1}{r^2}\right)$$

where we have written u for the retarded time $t - |\underline{x}|$, ω for the direction given by the unit vector $\frac{\underline{x}}{r}$, $r = |\underline{x}|$, and

$$(29)_2\qquad N_{ab}(u,\omega) \equiv \frac{1}{2\pi}\int T_{ab}\left(\underline{y}, u + \frac{\underline{x}\cdot\underline{y}}{r}\right) d^3y \quad .$$

From (29) it follows that

$$(30)_1\qquad \psi_{ab;c} = \frac{-\dot{N}_{ab}}{r}\, k_c + \theta\left(\frac{1}{r^2}\right)$$

$$(30)_2\qquad \psi_{ab;cd} = \frac{\ddot{N}_{ab}}{r}\, k_c k_d + \theta\left(\frac{1}{r^2}\right) \quad ,$$

here the dot indicates a partial derivative with respect to the retarded time for fixed ω, and k^a is chosen as in (16). Since the Einstein convention

(15) is satisfied in consequence of (13) we also have

$$(30)_3 \qquad \dot{N}_{ab}k^b = \theta(\frac{1}{r}) \quad , \quad \Rightarrow \ddot{N}_{ab}k^b = \theta(\frac{1}{r}) \quad .$$

$(30)_{2,3}$ give for the linearized curvature tensor the expression $\left\{ (23)_2 + \theta(\frac{1}{r^2}) \right\}$ with

$$(31) \qquad i_{ab} = (2r)^{-1} (\ddot{N}_{ab} - \frac{1}{2} \Delta_{ab} \ddot{N}^c{}_c) + \theta (\frac{1}{r^2}) \quad .$$

This result proves : The $\frac{1}{r}$- part of the curvature tensor which belongs to the retarded radiation field of a bounded source has the same algebraic structure as that of a plane wave.

The development indicated before $(29)_1$ can of course be carried on further ; the coefficients of the higher powers of r^{-1} will be functions of (u, ω) which can be represented by integrals like $(29)_2$ with T_{ab} replaced by its time derivatives, multiplied by polynomials in $\underline{y}^2$ and $\frac{\underline{x} \cdot \underline{y}}{r}$. We shall return to this result in the rigorous theory. If the changes within the source are sufficiently slow it is useful to develop the integrand of $(29)_2$ in powers of $\frac{\underline{x} \cdot \underline{y}}{r}$; this leads to a multipole expansion of the radiation field. Because of the energy-momentum conservation law (13) the lowest order radiation is of the quadrupole type.

We may finally ask : What is the energy carried away from a source by gravitational radiation? If we accept the gravitational energy tensor[20]

$$(32) \qquad t_{ab} = \frac{1}{4} (\psi_{cd,a} \bar{\psi}^{cd}{}_{,b} - \frac{1}{2} \psi_{,a} \psi_{,b} - \frac{1}{2} \Delta_{ab} \psi_{cd,e} \bar{\psi}^{cd,e} - \frac{1}{2} \psi_{,c} \psi^{,c})$$

which arises by linearisation of the Einstein energy-momentum affine tensor of the full theory and adding a term the divergence of which vanishes in consequence of (15) (and which can also be constructed by means of the Lagrangean formulation of the linear theory), we obtain from (15) and (16)

$$t_{ab} = \frac{1}{4}\left(\chi_{cd}\chi^{cd} - \frac{1}{2}(\chi^{c}{}_{c})^{2}\right) k_a k_b + \theta\left(\frac{1}{r^3}\right) . \tag{33}$$

The factor of $k_a k_b$ is never negative.

This expression (33) in conjunction with the expansion above can be and often has been used to calculate the gravitational energy loss of a spinning rod or double star etc. It can be used to estimate the radiation damping and the mass loss of such systems. In particular, one can specialize (28) to the case of a moving mass point (δ -like source along a world line) and treat the gravitational analogue of the Lienard-Wiechert potentials and fields well known in electrodynamics and then proceed to systems of a few mass points by superposition (simple quadrupoles etc). In view of the inconsistency of the linearized theory mentioned in the beginning of this section these considerations must be regarded with scepticism, however. A more satisfactory treatement has to include higher approximations such that the reaction of the field on the sources is accounted for. We shall not deal with this difficult question here.

The problems we want to consider in the rigorous theory are those of existence, propagation properties, action on test particles of free gravitational waves, and waves emitted by bounded sources, and we shall use the results of the linear theory as a guide.

3. The Petrov-classification of conformal curvature tensors

In section 1 we have seen that the quantity characterizing a gravitational field locally is the curvature tensor, and in section 2, that it is possible, at least in the linearized theory, to find a simple algebraic property, which holds rigourously for plane waves and asymptotically for waves emitted by a bounded source, namely eq. (27). It seems, therefore, useful to perform an algebraic classification of curvature tensors, try to define more or less "pure" radiation fields by weakening the condition (27) and then to look whether the different types of fields obey, rigorously or asymptotically for suitable boundary conditions, propagation laws generalizing eqs. (26) and (31), respectively. This idea has been originally put forward by Pirani and has been developed by Lichnerowicz, Bel, Sachs, Robinson, Trautman, Bondi, Penrose, Newman Unti, Tamburino, Kundt, myself, and others. It has proven to be very fruitful, at least if the differential-geometrical, strictly classical point of view towards general relativity is adopted, as I do in these lectures.

Let us now discuss the algebraic classification of curvature tensors of 4-dim., normal hyperbolic Riemannian manifolds.

The curvature tensor has the symmetry properties

$$(34) \qquad R_{abcd} = R_{[cd]ab}, \quad R_{[abcd]} = 0 \quad ;$$

a tensor obeying (34) , given at a point of a manifold, can always be realized as curvature tensor of a suitably chosen metric on this manifold. The linear space of tensors satisfying (34) is an irreducible representation space of the full linear group GL(R, 4).

With respect to the (homogeneous) Lorentz group this representation decomposes according to the formula

$$(35)_1 \qquad R_{abcd} = C_{abcd} + \frac{1}{12} R \; g_{abcd} + S_{abcd} \quad ,$$

here

$$(36) \qquad g_{abcd} = 2 \, g_{a\,[c}g_{d]\,b}$$

is the "bivector metric", and the tensors on the right hand side of $(35)_1$ are characterized by the symmetry properties (34) together with $(35)_1$ and their "dual symmetries"

$$(35)_2 \qquad {}^*S^* = S, \qquad {}^*C^* = - C$$

where we have omitted the indices. R is the curvature scalar as before.

We arrive at (35) by interpreting $R^{ab}{}_{cd}$ as a linear mapping of the bivector space $\{V^{cd}\}$ into itself and decomposing this mapping into that part, $C^{\cdot\cdot}{}_{\cdot\cdot} + \frac{1}{12} R \, g^{\cdot\cdot}{}_{\cdot\cdot}$, which commutes with the duality $V \rightarrow \overset{*}{V}$, and that part which anticommutes with it, which is called $S^{\cdot\cdot}{}_{\cdot\cdot}$. Finally, one decomposes the first mapping into a trace-free part, $C^{\cdot\cdot}{}_{\cdot\cdot}$, and a scalar multiplication, $\frac{R}{12} g^{\cdot\cdot}{}_{\cdot\cdot}$.

Explicit representations of C and S are

$$(37) \qquad C^{ab}{}_{cd} = R^{ab}{}_{cd} - 2\, \delta^{[a}_{[c} \; R^{b]}_{d]} + \frac{R}{6} \, \delta^{ab}_{cd} \quad ,$$

$$(38) \qquad S_{abcd} = - g_{abe\,[c} S^{e}{}_{d]}$$

$$(39) \qquad S_{ab} = R_{ab} - \frac{R}{4} \, g_{ab} \quad .$$

The tensors on the right hand side of $(35)_1$ have 10, 1, 9 linearly independent components, respectively. S_{abcd} is algebrically equivalent to S_{ab}, the trace-free part of the Ricci tensor. $C^a{}_{bcd}$ is Weyl's conformal curvature tensor, briefly called conform-tensor in th following, for which we will give a geometrical interpretation in the next section.

According to $(35)_1$ an algebraic classification of curvature tensors is obtained by first classifying Ricci tensors and conform tensors separately, and then taking the joint classification of both. Since we are mainly interested in vacuum fields where R..... reduces to C... and since, moreover R_{ab} within matter is expliciitely given if the stress energy momentum tensor is specified, we concentrate on the classification of conformtensors. (An analogous classification of R_{ab}'s has been given by Churchill).

The most elegant way to arrive at this classification, the Petrov-classification, makes use of the spinor calculus, as pointed out by Penrose. We shall follow his method since the spinor calculus is a very useful tool for the following investigations, too.

Let us remember the main definitions and relations of spinor algebra :

Let S be a complex, 2-dimensional vector space with a bilinear, alternating scalar product $[\varphi, \psi]$. Its element $\varphi, \psi, \ldots$ are called 2-contraspinors. The term "basis" shall be used only for pairs κ, μ of elements satisfying $[\kappa, \mu] = 1$. The components of $\varphi \subset S$ with respect to such bases are written φ^A, $\varphi^{A'}$ etc. as in tensor calculus; a change of the basis induces a unimodular transformation of the components. (We have a "symplectic geometry".) We write

$$(40) \qquad [\varphi, \psi] = \varepsilon_{AB}\, \varphi^A \psi^B \quad .$$

The components of the "metric spinor" are always given by

(41) $$\varepsilon_{(AB)} = 0 , \qquad \varepsilon_{12} = 1 .$$

We can now construct the dual space S^* of S ; its elements are called 2-cospinors and written φ_A ; the complex-conjugate space $\bar{S} = \{\varphi^{\dot{A}}\}$, and its dual $\bar{S}^* = \{\varphi_{\dot{A}}\}$. The metric allows to identify S and S^*, we write

(42) $$\varphi_B = \varphi^A \varepsilon_{AB} , \quad \varphi^A = \varepsilon^{AB} \varphi_B , \quad \varepsilon^{AB} \varepsilon_{CB} = \delta^A_C ,$$

and similar relations hold for $\bar{S}$ and $\bar{S}^*$, written with dotted indices, where

(43) $$\varepsilon_{\dot{A}\dot{B}} = \varepsilon_{AB} \text{ numerically if } \dot{A} = A , \ \dot{B} = B .$$

Obviously we can now introduce spinors with more indices and use the usual rules of tensor calculus. Care must be taken, however, with index-shiftings since ε_{AB} is skew; we have, e.g., $\varphi_A{}^A = - \varphi^A{}_A$.

We use the antilinear mapping

(44) $$\varphi^A \longrightarrow \bar{\varphi}^{\dot{A}} \equiv \overline{\varphi^A}$$

of S onto $\bar{S}$.

The order of indices of a different kind is immaterial, e.g. $\varphi^{A\dot{B}} = \varphi^{\dot{B}A}$.

The usefulness of this calculus for Minkowskian geometry is due to the following fact: Consider the vector spaces

(45) $$\tilde{V} \equiv \left\{\varphi^{A\dot{B}}\right\}, \qquad V \equiv \left\{\varphi^{AB} : \bar{\varphi}^{A\dot{B}} = \varphi^{A\dot{B}}\right\}$$

and the scalar product

(46) $$-\varphi_{A\dot{B}}\,\psi^{A\dot{B}} = -\,\varepsilon_{AB}\,\varepsilon_{\dot{C}\dot{D}}\,\varphi^{AC}\,\psi^{B\dot{D}}$$

which is real-valued over $V \times V$. Take a basis $\left\{\kappa^A, \mu^A\right\}$ of S,

(47) $$\kappa_A\,\mu^A = 1 \quad ,$$

and define

(48) $$t^{A\dot{B}} \equiv \mu^A\,\bar{\kappa}^{\dot{B}} \quad , \quad k^{A\dot{B}} \equiv \kappa^A\,\bar{\kappa}^{\dot{B}} \quad , \quad m^{A\dot{B}} = -\,\mu^A\,\bar{\mu}^{\dot{B}}$$

and (we omit indices)

(49) $$\underset{1}{e} \equiv \frac{1}{\sqrt{2}}\,(t + \bar{t}) \quad , \qquad \underset{2}{e} \equiv \frac{1}{i\sqrt{2}}\,(t - \bar{t}) \quad ,$$
$$\underset{3}{e} \equiv \frac{1}{\sqrt{2}}\,(k + m) \quad , \qquad \underset{4}{e} \equiv \frac{1}{\sqrt{2}}\,(k - m) \quad .$$

Then $\left\{t, \bar{t}, k, m\right\}$ is a basis of $\tilde{V}$, $\left\{\underset{a}{e}\right\}$ a basis of V, and the scalar products are, according to (46), (47), (48), and (49),

$(50)_1$ $$t \quad t = k \cdot k = m \cdot m = t \cdot k = t \cdot m = 0, \quad t \cdot \bar{t} = k \cdot m = 1 \; ,$$

$(51)_1$ $$\underset{a}{e}\;\underset{b}{e} = \Delta_{ab}$$

The metric $(46)_1$ in V has, accordingly, the signature +++- ; V is a Minkowski-space.

We can describe the elements of $\tilde{V}$ (or V) by means of their components φ^a with respect to an arbitrary basis of $\tilde{V}$ (V) instead of using the "spinor" components" $\varphi^{A\dot{B}}$, and then write

(52) $$\varphi^{A\dot{B}} = \varphi^a \sigma_a{}^{A\dot{B}}, \quad \varphi^a = - \sigma^a{}_{A\dot{B}} \varphi^{A\dot{B}} .$$

A change $\{\kappa^A, \mu^A\} \longrightarrow \{'\kappa^A, '\mu^A\}$ of the basis of S induces, via (49), a Lorentz-transformation in V, and one can prove that this homomorphism of the unimodular group SL(C, 2) ≡ $\mathfrak{g}$ into the homogeneous Lorentz group is, in fact, a covering homomorphism of $\mathfrak{g}$ onto $L_+^\uparrow$, the proper, orthochronous part (identity component) of that group. The kernel of this homomorphism consists of 1 and -1 only :

(53) $$\mathfrak{g}/\{\pm 1\} \cong L_+^\uparrow$$

Since $\mathfrak{g}$ is simply connected (in contradistinction from $L_+^\uparrow$ which is doubly connected), $\mathfrak{g}$ is the universal covering group of $L_+^\uparrow$.

The construction given above is nothing than a geometrical disguise of the algebraic relation (53) : An oriented and time-oriented Minkowski (vector) space V may be considered as constructed by means of a spin-space S as described above; (48), (49) establish a two to one-relationship

(54) $$\pm \{\kappa, \mu\} \longleftrightarrow \{e_a\}$$

between the (unimodular) bases of S and the orthonormal, oriented and time oriented bases of V.

It is now obvious that we may identify the tensors associated with V with certain spinors associated with S, the components being related by formulae like (52); $\sigma^{\cdot}{}_{\cdot\cdot}$ may be considered as a connecting quantity in the sense of Schouten. We shall write, e.g.

(55) $$F_{ab} \leftrightarrow F_{A\dot{C}B\dot{D}} = F_{ab}\,\sigma^{a}{}_{A\dot{C}}\,\sigma^{b}{}_{B\dot{D}} ,$$

i.e. the "kernel" F is left unchanged in passing from tensor to spinor components and vice versa. Algebraic relations can easily be translated from tensor to spinor notation; each contraction over a pair of tensor indices gives rises, according to (52), to a factor -1 in front of the contraction over the corresponding pair of spinor indices. (This is done in order to have the signature +++-, and not ---+).

One of the main reason for the usefulness of the spinor calculus is the fact that the irreducible representation of $\mathcal{L}$ are given by spinors completely symmetric in the undotted and the dotted indices, The decomposition into irreducible parts can be performed according to the following scheme[26]:

$$0 = \varepsilon_{A[B}\,\varepsilon_{CD]}\,\varphi^{CD} = \varepsilon_{AB}\,\varphi_{D}{}^{D} - 2\,\varphi_{[AB]} ,$$

(56)

$$\text{i.e.}\quad \varphi_{[AB]\ldots} = \frac{1}{2}\,\varepsilon_{AB}\,\varphi_{D}{}^{D}{}_{\ldots} ,$$

$$\psi_{(ABC)} = \frac{1}{3}\left(\psi_{A(BC)} + \psi_{B(CA)} + \psi_{C(AB)}\right)$$

$$= \psi_{A(BC)} + \frac{1}{3}\left(\psi_{B(CA)} - \psi_{A(CB)}\right) + \frac{1}{3}\left(\psi_{C(AB)} - \psi_{A(CB)}\right)$$

$$= \psi_{ABC} - \frac{1}{2}\,\varepsilon_{BC}\,\psi_{AD}{}^{D} + \frac{1}{3}\left(\varepsilon_{AB}\,\psi^{D}{}_{(DC)} + \varepsilon_{AC}\,\psi^{D}{}_{(DB)}\right) ,$$

(57) i.e. $\psi_{ABC} = \psi_{(ABC)} + \frac{1}{2}\,\varepsilon_{BC}\,\psi_{AD}{}^{D} - \frac{1}{3}(\,\varepsilon_{AB}\,\psi^{D}{}_{(DC)} + \varepsilon_{AC}\,\psi^{D}{}_{(DB)}$

which is the desired decomposition of ψ_{ABC} into irreducibile parts corresponding to the Clebsch-Gordan-formula

$$D_{\frac{1}{2}} \times D_{\frac{1}{2}} \times D_{\frac{1}{2}} = (D_1 + D_o) \times D_{\frac{1}{2}} = D_{\frac{3}{2}} + 2D_{\frac{1}{2}} \quad .$$

We recognize : Any spinor can be written as a sum of terms consisting of ε-factors and totally symmetric spinors, the latter arising from the former by taking symmetric parts and traces.

A (real) vector k^a corresponds to a Hermitian spinor $k^{A\dot{B}}$, and a future-pointing null vector, in particular, to a spinor of the form $k^{A\dot{B}} = \kappa^A \bar{\kappa}^{\dot{B}}$ as is obvious from the identity $\kappa_A \kappa^A = 0$ and the transitivity of the Lorentz group with respect to the nulle cone.

A direction on the light cone is given by a "spinray" $\{\lambda\kappa^A\}$, $0 \neq \lambda \in C$, i.e. by a complex number $\frac{\kappa^2}{\kappa^1}$ (∞ being admitted). Lorentz transformations of these null directions correspond in a one to one way to transformations $z \rightarrow \frac{az + b}{cz + d}$ of the "Argand" or "Riemann" sphere of these numbers , which is another way of stating the local isomorphism (53) . From this correspondence we infer that $L_{+}^{\uparrow}$ acts on the set of null directions in a threefold transitive way.

The basis transformations

$$(58) \qquad {}'\kappa^A = A\,\kappa^A \,, \quad {}'\mu^A = A^{-1}\,\mu^A \,, \qquad A \in C$$

represent space like rotations if $|A| = 1$, time-like rotations if $A \in R$. The transformations

(59) $\quad 'x^A = x^A , \quad '\mu^A = \mu^A + B x^A , \quad B \in C$

are called null rotations.

Real bivectors F_{ab} may be decomposed, according to the scheme (56), as

(60) $$F_{ab} \longleftrightarrow F_{A\dot{C}B\dot{D}} = \frac{1}{2}(\phi_{AB}\, \varepsilon_{\dot{C}\dot{D}} + \varepsilon_{AB}\, \bar{\phi}_{\dot{C}\dot{D}}) ,$$

$$\phi_{AB} = F_{A\dot{C}B}{}^{\dot{C}} = \phi_{BA} ,$$

i.e. they are represented by "bispinors" ϕ_{AB}.

If the "translation" (55) is applied to the curvature tensor, a spinor $R_{A\dot{E}\ B\dot{F}\ C\dot{G}\ D\dot{H}}$ is obtained. It may be decomposed into irreducibile parts by the method described above; then the spinor analogue of $(35)_1$ results. I only write down the decomposition of the conform tensor:

(62) $$C_{abcd} \longleftrightarrow C_{A\dot{E}B\dot{F}C\dot{G}D\dot{H}} = \frac{1}{2} (\Gamma_{ABCD}\, \varepsilon_{\dot{E}\dot{F}}\, \varepsilon_{\dot{G}\dot{H}} + hc) ,$$

here "h.c." denotes the Hermitian conjugate as in (60).

The converse of (62) is given by

(63) $$\Gamma_{ABCD} = \frac{1}{2}\, C_{A\dot{E}B}{}^{\dot{E}}{}_{C\dot{G}D}{}^{\dot{G}} .$$

The complicated symmetry-and duality properties of C_{abcd} are equivalent with the statement that a representation (62) holds and Γ_{ABCD}, the conformspinor, is totally symmetric.

Now it is easy to classify the C' s by classifying the Γ's , to obtain normal forms, and to give geometrical interpretations of the results.

According to the fundamental theorem of algebra the binary quartic $\Gamma(\xi) \equiv \Gamma_{ABCD}\,\xi^A \xi^B \xi^C \xi^D$ is a product of linear factors. That means there exist spinors $\overset{i}{\varkappa}_A$ such that

(64) $$\Gamma_{ABCD} = \overset{1}{\varkappa}_{(A}\overset{2}{\varkappa}_B\overset{3}{\varkappa}_C\overset{4}{\varkappa}_{D)} ,$$

the corresponding null directions being determined by

(65) $$\Gamma(\overset{i}{\varkappa}) = 0 .$$

Naturally one may assign multiplicities to these "principal null directions" of C... Let us write C : 2, 1, 1 to express that C has one 2fold and two simple p.n.d., etc. Then we may define the Petrov-types conformtensors (-spinors) by the following table :

(66)

I : 1, 1, 1, 1; II: 2, 1, 1; D : 2, 2;

III : 3, 1; N : 4; 0: --- .

The last type, 0 , consists of the zero conform tensor only; this type belongs, as is well-konwn, to conformally flat spaces. The symbols I, ..., 0 are to be used as names of the types and also as kernel letters of C...'s of Γ..'s belonging to these types.

This classification is not only invariant under Lorentz-transformations but also under "duality rotations" given by

$(67)_1$ $$\Gamma\ldots \longrightarrow e^{-i\theta}\,\Gamma\ldots$$

or, tensorially, by

(67)$_2$ $$C\ldots \rightarrow \cos\theta\, C\ldots + \sin\theta\, {}^*C\ldots ,$$

and under moltiplications with real scalars.

One can now use the fact that triples of null directions can be transformed into arbitrary positions on the Argand sphere (see above) in order to obtain normal forms for the "special" types, i.e. those admitting at least one degenerate null direction. Such normal forms are

(68) $$N_{ABCD} = \kappa_A \kappa_B \kappa_C \kappa_D ,$$

(69) $$III_{ABCD} = \kappa_A \kappa_B \kappa_{(C} \mu_{D)} + \kappa_{(A} \mu_{B)} \kappa_C \kappa_D ,$$

(70) $$D_{ABCD} = -\lambda (6 \kappa_{(A} \mu_{B)} \kappa_{(C} \mu_{D)} - \varepsilon_{A(C} \varepsilon_{D)B}) , \quad = 0 ,$$

(71) $$II_{ABCD} = -\lambda (4 \kappa_{(A} \mu_{B)} \kappa_{(C} \mu_{D)} + \kappa_A \kappa_B \mu_C \mu_D + \mu_A \mu_B \kappa_C \kappa_D) + 4 \kappa_A \kappa_B \kappa_C \kappa_D , \quad \lambda \neq 0 ,$$

here is always

(72) $$\kappa_A \kappa^A = 1 .$$

In the cases II and III the "eigenbasis" $\{\kappa_A, \mu_A\}$ is fixed up to a finite ambiguity, in case N up to null rotations, and in case D up to rotations (58).

A non-special conformspinor can be written (in three different ways) as

$$I_{ABCD} = -\lambda\,(6\,\kappa_{(A}\mu_{B)}\kappa_{(C}\mu_{D)} - \varepsilon_{A(C}\varepsilon_{D)B}) \qquad (73)$$
$$+\gamma\,(\kappa_A\kappa_B\kappa_C\kappa_D + \mu_A\mu_B\mu_C\mu_D)\,,$$
$$\neq 0$$

where again (72) holds.

These normal forms can, of course, be written in tensor form, but the resulting formulae are complicated, and we shall not need them.

We want to mention, however, the equations which characterize a spinor κ^A as a simple, double, etc. eigenspinor of Γ_{ABCD}, and we also write down the corresponding tensor equations for C_{abcd} and $k^a \leftrightarrow \kappa^A\bar{\kappa}^{\dot{B}}$:

$$\kappa^A \text{ simple}: \quad \Gamma_{ABCD}\,\kappa^A\kappa^B\kappa^C\kappa^D = 0$$
$$\Leftrightarrow k_{[a}C_{b]cd[e}k_{f]}k^c k^d = 0\,,$$
$$\kappa^A \text{ twofold}: \quad \Gamma_{ABCD}\,\kappa^B\kappa^C\kappa^D = 0$$
$$\Leftrightarrow C_{bcd[e}k_{f]}k^c k^d = 0\,, \qquad (74)$$
$$\kappa^A \text{ threefold}: \quad \Gamma_{ABCD}\,\kappa^C\kappa^D = 0\,,$$
$$\Leftrightarrow C_{bcd[e}k_{f]}k^d = 0\,,$$
$$\kappa^A \text{ fourfold}: \quad \Gamma_{ABCD}\,\kappa^D = 0$$
$$\Leftrightarrow C_{abcd}k^d = 0\,.$$

We shall not present the proofs since these formulae are almost immediate consequences of the definitions involved.

We now see, however, that the null type is identical with the pure radiation type to which we were led in the linear approximation. We may arrange the types in the Penrose-diagramm

(75)
$$\begin{array}{ccccc} & & I & & \\ & \swarrow & & \searrow & \\ & II & \longrightarrow & D & \\ \swarrow & & \searrow & & \searrow \\ III & \longrightarrow & N & \longrightarrow & O \end{array}$$

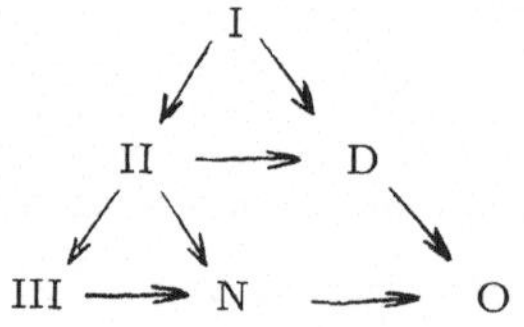

in which arrows denote specialisations in an obvious way. We will conjecture, therefore, that the types N, III, D, II in this orther will describe less and less "pure" gravitational waves, i.e. waves far away from matter, and therefore next ask whether these types have propagation properties.

We finally remember that the algebraic structure of $C...$, that is of R_{abcd} in empty space, admits an interpretation in terms of test body behaviour which we have described explicitely in section II for - as we may now say - the null type.

4. Spinor-differential-calculus

Before studying the propagation of curvature tensors let us extend the spinor calculus to include analysis. I shall develop the spinor analysis as an exterior differential calculus here analogously to the well-known Cartan-calculus since this seems to be the most compact and effective formalism available. The calculus to be presented is due to K. Bichteler.

As a preliminary, I review very briefly the Cartan procedure. One writes for the general infinitesimal displacement

(76) $$\vec{dx} = \theta^a \vec{e}_a$$

where $\{\vec{e}_a\}$ is an arbitrary tetrad field on the space time manifold W. The Pfaffians θ^a form what is called a vector 1-form; the metric is given by

(77) $$\vec{dx}^2 = g_{ab}\, \theta^a \theta^b ,$$

Next, one introduces a covariant differential D defined on the set of tensor-r-forms (i.e. tensor-valued alternating differential forms of degree r):

(78) $$D\, T^{ab}{}_c = dT^{ab}{}_c + \omega^a{}_d \wedge T^{db}{}_c + \omega^b{}_d \wedge T^{ad}{}_c - \omega^d{}_c \wedge T^{ab}{}_d$$

satisfying the usual rules.

The connection is required to have vanishing torsion,

(79) $$D\, \theta^a = 0 ,$$

and to be metrical,

(80) $$D\, g_{ab} = 0 ,$$

and is therefore uniquely fixed by the metric.

The curvature form Ω is introduced through

(81) $$D^2 T^a{}_b = \Omega^a{}_c \wedge T^c{}_b - \Omega^c{}_b \wedge T^a{}_c ,$$

and the curvature tensor is introduced through the development

(82) $$\Omega^a{}_b = -\frac{1}{2} R^a{}_{bcd}\, \theta^c \wedge \theta^d \quad , \qquad R^a{}_{b(cd)} = 0 \quad .$$

How can this be imitated for spinors ? First, one introduces a spin-structure on a manifold by associating with each point x a spin space S_x ; one constructs, according to the last section, from S_x the space V_x and, finally, identifies V_x with the tangent space of W at x such that the connecting quantity $\sigma^a{}_{AB}(x)$ is differentiable. The latter requirement can always be satisfied if the manifold is orientable and time-orientable; the following formalism can therefore be used locally without restriction of generality.

Let, then, $\{\kappa(x), \mu(x)\}$ be a basis of S_x and let the Pfaffian forms $\theta^{A\dot{B}}$ be the spinor components of dx ,

(83) $$\theta^{A\dot{B}} = dx^a\, \sigma_a{}^{A\dot{B}} \quad .$$

$\theta^{A\dot{B}}$ is an example of a spinor-1-form. The metric is then expressed by

$$\vec{dx}^2 = -\theta_{A\dot{B}}\theta^{A\dot{B}}$$

in line with (46) .

We introduce a bispinor-2-form by

(84) $$\theta^{AB} \equiv \frac{1}{2}\, \theta^A{}_{\dot{C}} \wedge \theta^{B\dot{C}}$$

and a real scalar-4-form θ by

(85) $$\theta^{A\dot{P}} \wedge \theta^{B\dot{Q}} \wedge \theta^{C\dot{R}} \wedge \theta^{D\dot{S}} = \theta\, \eta^{A\dot{P}B\dot{Q}C\dot{R}D\dot{S}} \quad .$$

From the spinor-representation of η ....,

$$\eta^{A\dot{B}C\dot{D}}{}_{E\dot{F}G\dot{H}} = i(\delta^{A}_{E}\delta^{C}_{G}\delta^{\dot{B}}_{\dot{A}}\delta^{\dot{D}}_{\dot{F}} - \delta^{A}_{G}\delta^{C}_{E}\delta^{\dot{B}}_{\dot{F}}\delta^{\dot{D}}_{\dot{H}}) \tag{86}$$

follow the relations

$$\theta^{AB} \wedge \theta^{CD} = i\theta\, \varepsilon^{A(C}\varepsilon^{D)B}, \quad \theta^{AB} \wedge \bar{\theta}^{\dot{C}\dot{D}} = 0, \tag{87}$$

$$\theta = \frac{i}{3}\, \theta_{AB} \wedge \theta^{AB}$$

Next, we define a spin-affine-connection or covariant differential for spinor r-forms, e.g,

$$D\,\phi^{AB\dot{C}} = d\,\phi^{AB\dot{C}} + \omega^{A}{}_{D} \wedge \phi^{DB\dot{C}} + \Omega^{B}{}_{D} \wedge \phi^{AD\dot{C}} + \bar{\Omega}^{\dot{C}}{}_{\dot{D}} \wedge \phi^{AB\dot{D}} \tag{88}$$

Here $(\omega^{A}{}_{B})$ is a matrix of Pfaffians, the "connection form".

This mapping D satisfies the usual rules of a derivation : Sum-and (Leibniz) product rule, commutativity with contractions, $D\varphi = d\varphi$ for scalars φ, and "reality", i.e. commutativity with the conjugation operation (44).

To relate the D-operator to the Riemannian metric, we require vanishing torsion,

$$D\,\theta^{A\dot{P}} = 0 \qquad (\Longleftrightarrow D\,\theta^{AB} = 0), \tag{89}$$

and the validity of the "Ricci lemma" for the spin-metric,

(90) $$D\,\varepsilon_{AB} = 0 \;(\Longleftrightarrow \omega_{[AB]} = 0)\;.$$

The connection $\omega^A{}_B$ is uniquely determined by the last two requirements; the Pfaffians $\omega^A{}_B$ can be calculated from

(91) $$d\,\Theta^{AB} + \omega^A{}_C \wedge \Theta^{CB} + \omega^B{}_C \wedge \Theta^{AC} = 0,\quad \omega^A{}_A = 0$$

which is just a restatement of the two preceding equations.

Just as in the tensorial formulation, the operator D^2 gives rise to the introduction of a curvature 2-form, e.g.

$(92)_1$ $$D^2 \Phi^A{}_B = \Omega^A{}_C \wedge \Phi^C{}_B - \Omega^C{}_B \wedge \Phi^A{}_C\,,\quad \Omega^A{}_A = 0\,,$$

here

$(92)_2$ $$\Omega^A{}_B = d\omega^A{}_B + \omega^A{}_C \wedge \omega^C{}_B\;.$$

Let us apply the operator D to the "Ricci identity" $D^2 \xi^A = \Omega^A{}_B\, \xi^B$. We obtain $D^3 \xi^A = D\Omega^A{}_B\, \xi^B + \Omega^A{}_B \wedge D\xi^B = D^2(D\xi^A) = \Omega^A{}_B \wedge D\xi^B$, consequently

(93) $$D\,\Omega^A{}_B = 0\;.$$

This is the Bianchi-identity, written in terms of the spincurvature-form.

According to the definition (83) the components of $\Theta^{A\dot{B}} \wedge \Theta^{C\dot{D}}$ form a basis of the set of 2-forms. Due to the bivector decomposition (60) the same is true for the components of Θ^{AB} together with $\bar{\Theta}^{\dot{A}\dot{B}}$. We may therefore develop the curvature form:

$$\Omega_{AB} = -\frac{1}{2}\left(\Gamma_{ABCD} - \frac{R}{6}\varepsilon_{A(C}\varepsilon_{D)B}\right)\Theta^{CD} - \frac{1}{2}S_{AB\dot{P}\dot{Q}}\bar{\Theta}^{\dot{P}\dot{Q}} . \tag{94}$$

The spinors $\Gamma_{\ldots}$, $S_{\ldots}$ and the scalar R which appear as coefficients are, in fact, the conformspinor, the spinor corresponding to the reduced Ricci tensor, and the curvature scalar, respectively. This can be seen as follows:

According to (81) we have $D^2k^a = \Omega^a{}_b k^b$ for an arbitrary vector k^a. In spinor notation this reads $D^2k^{A\dot{E}} = -\Omega^{A\dot{E}}{}_{B\dot{F}}k^{B\dot{F}}$. On the other hand, $(92)_1$ gives $D^2k^{A\dot{E}} = (\Omega^A{}_B\delta^{\dot{E}}_{\dot{F}} + \bar{\Omega}^{\dot{E}}_{\dot{F}}\delta^A_B)k^{B\dot{E}}$. Comparison yields

$$\Omega^A{}_B = -\frac{1}{2}\Omega^{A\dot{E}}{}_{B\dot{E}} . \tag{95}$$

Finally, (82) gives

$$\begin{aligned}\Omega^A{}_B &= -\frac{1}{2}R^{A\dot{E}}{}_{B\dot{E}} = \frac{1}{2}R^{A\dot{E}}{}_{B\dot{E}C\dot{G}D\dot{H}}\Theta^{C\dot{G}}\wedge\Theta^{D\dot{H}} \\ &= -\frac{1}{4}R^{A\dot{E}}{}_{B\dot{E}C\dot{G}D\dot{H}}(\Theta^{CD}\varepsilon^{\dot{G}\dot{H}} + \varepsilon^{CD}\bar{\Theta}^{\dot{G}\dot{H}}) ,\end{aligned} \tag{96}$$

and if we put equal the coefficients in (94) and (96) the resulting relations are identical with those used in the algebraic decomposition of the Riemannspinor.

From (93), (94) and (89) the Bianchi identities in terms of the spinors Γ_{ABCD}, $S_{AB\dot{C}\dot{D}}$, and R can be obtained. We write down the Bianchi-identity for a vacuum field only; it is

(97) $$\Gamma_{ABCD;}{}^{D\dot{E}} = 0 .$$

The covariant derivative is defined, of course, by $D\,\Gamma_{ABCD} \equiv -\Gamma_{ABCD;E\dot{F}}\,\theta^{E\dot{F}}$

If the Ricci-identity is written explicitely in terms of components of $D^2\,\xi^A$, $\Omega^A{}_B$ etc. we obtain, among other relations, in the vacuum case

(98) $$\xi_{A;B(\dot{C}}{}^{B}{}_{\dot{D})} = 0.$$

The connection form $\omega^A{}_B$ can also be developed :

(99) $$\omega^A{}_B = -\Gamma^A{}_{BC\dot{P}}\,\theta^{C\dot{P}} , \Gamma^A{}_{AC\dot{P}} = 0 .$$

Since the components of the basis-spinors κ^A, μ^A are constants, we have, e.g. $D\,\kappa_A = -\kappa_{A;C\dot{P}}\,\theta^{C\dot{P}} = -\omega^B{}_A\,\kappa_B = -\Gamma_{ABC\dot{P}}\,\kappa^B\,\theta^{C\dot{P}}$, i.e.

(100) $$\Gamma_{A1C\dot{P}} \overset{*}{=} \kappa_{A;C\dot{P}} ,$$

and a similar realtion for μ_A .

The star indicates that components with respect to $\{\kappa, \mu\}$ are to be used. The twelve complex quantities $\Gamma_{\dots}$ are the spinor-analogues of the 24 real Ricci rotation coefficients. Following Newman and Penrose we call them spin-coefficients. These spin-coefficients satisfy a system of 18 first order differential equations which are useful for the study of the propagation of gravitational waves. They are obtained by inserting (99) into the formula $(92)_2$ for the curvature form, developing the resulting expression with respect to θ^{AB} and $\bar{\theta}^{\dot{A}\dot{B}}$, and equating it to the right hand side of eq. (94).

During this calculation the Pfaffian derivatives of the Γ's, defined by

$$(101) \qquad d\,\Gamma_{AB C\dot{D}} \equiv -\,\Gamma_{ABC\dot{D}/E\dot{F}}\;\Theta^{E\dot{F}} \quad ,$$

are introduced. According to the definition of $\Theta^{A\dot{B}}$ we have, e.g.

$$(102) \qquad -\,\Gamma_{ABC\dot{D}/1\dot{1}} = \Gamma_{ABC\dot{D},a}\,k^{a} \quad ,$$

which is the derivative in the ray direction k^a. The "propagation equations" derived in this way relate differences of directional derivatives of the spin coefficients to bilinear expressions in the spincoefficients and their complex conjugates, and to components of the curvature spinor.

5. Congruences of null-geodesics

The considerations on plane waves and waves from bounded sources in the linear approximation and also the existence of preferred null vector fields - given by the principal null directions of the conformtensor - in a general space time suggest to investigate congruences of null-lines, in particular geodesics.

Let $x^a = x^a(y^\alpha, v)$ be a parameter representation of a congruence of null lines, y^α being parameters distinguishing the curves and v being an arbitrary parameter along each curve. We then have

$$(103) \qquad k^a \equiv \frac{\partial x^a}{\partial v} \quad , \qquad k_a k^a = 0 \; .$$

In order to study geometrical properties of this congruence we introduce a spinor basis $\{\chi, \mu\}$ such that

(104) $$k^a \longleftrightarrow k^{A\dot{B}} = \varkappa^A \bar{\varkappa}^{\dot{B}} .$$

$\{\varkappa, \mu\}$ define via eqs. (48), (49) a quasi-orthogonal tetrad field $\{t, \bar{t}, k, m\}$ and an orthonormal frame $\{\underset{a}{e}\}$. These frames are determined by the congruence only up to transformations (58), (59).

The lines of the congruence are geodesics if and only if $k_{[a}k_{b];c}k^c = 0$. In spinor language this condition reads

(105) $$\varkappa_{A;B\dot{C}} \varkappa^A \varkappa^B \bar{\varkappa}^{\dot{C}} = 0 .$$

In the geodesic case we may choose v to be an affine parameter, $k_{a;b}k^b = 0$. We may then take $\varkappa^A$ to be parallely propagated along the rays, too :

(106) $$\varkappa_{A;B\dot{C}} \varkappa^B \bar{\varkappa}^{\dot{C}} = 0 .$$

Let us restrict the following considerations to geodesic null congruences and take v as an affine parameter always.

The connection vector

(107) $$\delta x^a = \frac{\partial x^a}{\partial y^\alpha} \delta y^\alpha$$

of two neighbouring "rays" is Lie-propagated along the initial ray,

(108) $$\delta x^a{}_{;b} k^b = k^a{}_{;b} \delta x^b .$$

This equation follows immediately from $\frac{\partial^2 x^a}{\partial v \partial y^\alpha} = \frac{\partial^2 x^a}{\partial y^\alpha \partial v}$ if a geodesic coordinate system is used.

Eq. (108) implies that $k_a \delta x^a$ is constant along a ray. Let us call δx^a "transverse" if

$$k_a \delta x^a = 0 . \tag{109}$$

This property is independent of the parameterisation since it is not changed under the a substitution $\delta x^a \longrightarrow \delta x^a + \lambda k^a$, and it is propagated along the rays according to the remark preceding eq. (109). We may therefore speak of transversely connected adjacent rays.

The scalar product $\delta x^a \delta' x_a$ of two connection vectors satisfying (109) is independent of the parameterisation. It is, therefore, meaningful to speak of distances between transversely connected neighbouring rays and of angles between directions pointing from a ray L to two neighbouring rays L', L'' transversely connected with L. These geometrical quantities will change along L, and these changes are the geometrical properties we want to study.

For this purpose we use the bases introduced above and write

$$\delta_\perp x^a = x \underset{1}{e}^a + y \underset{2}{e}^a = \sqrt{2}\ R(\xi \bar{t}^a) , \tag{110$_1$}$$

$$x + iy \equiv \xi . \tag{110$_2$}$$

$\delta_\perp x^a$ is the projection of δx^a into the $\underset{1}{e}^a$, $\underset{2}{e}^a$-plane which is transverse in the sense of (109) and may be visualised as a screen which is traversed by the rays.

Let us momentarily choose $\{x, \mu\}$ such that (106) and

(111) $$\mu_{A;B\dot{C}}\,\varkappa^{B}\,\bar{\varkappa}^{\dot{C}} = 0$$

hold; then t, k, m, e_a are all constant along the rays. We can now calculate, by means of (108), the change of ξ defined in (110) for a connection vector satisfying (109). The result is

(112) $$\frac{d\xi}{dv} = \bar{z}\,\xi + \bar{\sigma}\,\bar{\xi} \quad ,$$

here

(113) $$z \equiv k_{a;b}\,\bar{t}^{a}t^{b} \quad , \quad \sigma \equiv k_{a;b}\,\bar{t}^{a}\bar{t}^{b}$$

are two complex numbers characterising the transverse derivatives of k^a. Eq. (112) describes the mapping of a "screen" at v onto a screen at $v+dv$ produced by the rays ; ξ is the "position vector" in the first, $\xi + d\xi$ the one ine the second screen.

Let us write

(114) $$z = \Theta + i\omega \quad .$$

Then we may interprete (112) : The infinitesimal transformation $\xi \longrightarrow \xi + d\xi$ consists of a rotation through the angle ωdv , an isotropic expansion of magnitude $\Theta\, dv$, and an area-preserving deformation of magnitude $|\sigma|$ the principal axes of which are determined by the real and imaginary parts of that vector t^a for which σ is real. (This latter condition can be satisfied by a spatial rotation $t \longrightarrow e^{i\theta}\, t$.)

According to our remarks above the differentials

$$z \quad dv, \quad |\sigma| \, dv$$

are invariants of the congruence, i. e. independent of the parameterisation, and z and σ may be calculated from (113) whether or not the condition $t_{a;b}k^b = 0$ holds.

We call θ the expansion, ω the twist, and σ the shear of the congruence with respect to v.

In spinor notation, (113) read

(115) $z = \varkappa_{A;B\dot{D}} \varkappa^A \mu^B \bar{\varkappa}^{\dot{D}}$, $\qquad \sigma = \varkappa_{A;B\dot{D}} \varkappa^A \varkappa^B \bar{\mu}^{\dot{D}}$.

From (105) and (115) we see: A null congruence is geodesic and shear free ($\sigma = 0$) if and only if (in an arbitrary parameterisation)

(116) $$\varkappa_{A;B\dot{C}} \varkappa^A \varkappa^B = 0$$

If this condition is satisfied we may choose $\varkappa^A$ such that the stronger condition

(117) $$\varkappa_{A;B\dot{D}} \varkappa^B = \alpha \varkappa_A \bar{\varkappa}_{\dot{D}}.$$

(which implies (106)) holds.

Comparing (115) with (100) we see that z and σ are certain spin coefficients associated with the spin basis $\{\varkappa, \mu\}$ adapted to the congruence. We have, in fact,

(118) $$z = \Gamma_{112\dot{1}}, \qquad \sigma = \Gamma_{1112}.$$

It is possible to give geometrical interpretations of further spin coefficients in terms of the congruence given by $\varkappa^A$, but we shall not discuss them here.

6. Geometry of principal null congruences of vacuum fields

We have now assembled the tools necessary for studying the null congruences defined by the principal null directions of the conformtensor of a vacuum field.

The first interesting theorem states that the congruence determined by a multiple principal null direction of a vacuum field is geodesic and shearfree.

To prove this theorem we use the equation $(74)_2$ which characterises $\varkappa^A$ to be a multiple p. n. d. , and the vacuum Bhianchi-identity (97) . According to these equations we have $0 = \varkappa^A \varkappa^B \varkappa^C \Gamma_{ABCD;}{}^{D\dot{E}} =$

$= -3\, \varkappa^A \varkappa^B \Gamma_{ABCD} \varkappa^{C;D\dot{E}}$. If $\varkappa^A$ is a twofold principal null direction we have, according to (74), $\Gamma_{ABCD} \varkappa^A \varkappa^B = \alpha \varkappa_C \varkappa_D$ with $\alpha \neq 0$, consequently eq. (114) is satisfied, and the proof is finished. If, however, $\varkappa^A$ has multiplicity 3 , we use analogously $0 = \varkappa^B \varkappa^C \Gamma_{ABCD;}{}^{D\dot{E}} =$

$= -2\, \varkappa^B \Gamma_{ABCD} \varkappa^{C;D\dot{E}} = -2\alpha\, \varkappa_B \varkappa_C \varkappa_D \varkappa^{C;D\dot{E}}$, $\alpha \neq 0$, and similarly in the case of a fourfold null direction.

Golberg and Sachs have shown that the converse is also true: If a vacuum field admits a shearfree, geodesic null congruence, its curvature tensor is algebraically special, the tangent of the congruence being a multiple principal null direction of it. The proof of this theorem can be given, with the formalism presented here, by combining some of the Bianchi-identities (97) with some of the "propagation equations" described at the end of section 4, namely those which contain the relevant components Γ_{A111} of the conformspinor only. We shall not carry out this proof here.

The two preceding theorems give a differential geometrical interpretation of the multiple p. n. d. of special Einstein spaces, and they also support the idea that these space times represent pure radiation since they define "rays" which have properties analogous to those studied in geometrical optics.

7. Propagation of curvature tensors along null congruences

Let us assume that we have a vacuum null field. We may then write, according to the preceding considerations,

$$\Gamma_{ABCD} = \nu \varkappa_A \varkappa_B \varkappa_C \varkappa_D , \tag{119}$$

$$\varkappa_A \mu^A = 1, \qquad \dot{\varkappa}_A = \dot{\mu}_A = 0 . \tag{120}$$

The dot indicates the covariant derivative in the ray direction $\varkappa^A \bar{\varkappa}^{\dot{B}}$.

From these equations and the Bianchi identity we obtain

$$0 = \mu^A \mu^B \mu^C \Gamma_{ABCD;}{}^{D\dot{E}} \bar{\varkappa}_{\dot{E}} = (\nu \varkappa_D)_;{}^{D\dot{E}} \bar{\varkappa}_{\dot{E}} \tag{121}$$

$$= - \dot{\nu} - \nu z$$

if we use the gage (115) of $\varkappa_A$ and the formula (115) for the complex expansion z of the $\varkappa^A$-congruence.

Since z is one of the spin coefficients - see (116) - and $\sigma = 0$ and $R_{ab} = 0$, one of the propagation equations reduces to the simple form

$$\dot{z} + z^2 = 0 . \tag{122}$$

The last two equations imply: Along a ray, either $z = 0$ and $\nu = \text{const.}$, or $z \neq 0$ and $\frac{\nu}{z} = \text{const.}$ Because of (119), (120), Γ_{ABCD} is either constant along a ray or differs from the constant null conformspinor $\varkappa_A \varkappa_B \varkappa_C \varkappa_D$ by a factor const. z only. This behaviour has a simple meaning : Let us define a "luminosity-distance" r along the rays by $\Theta = \frac{\dot{r}}{r}$ meaning that r^2 behaves like the cross section of a bundle rays, and a "twist angle" φ by $\omega = \dot{\varphi}$ such that $z = (\log r + i\varphi)^{\cdot}$. Then the solution of (122) can be written

$$z = r^{-1} e^{-i\varphi} \quad , \tag{123}$$

and we see that if the congruence expands, $\Theta \neq 0$, the magnitude of the conformtensor is inversely proportional to the luminosity distance, and the directional properties of this tensor change according to a duality rotation the angular velocity of which is equal to the twist velocity of the rays.

The preceding considerations can be generalized without much difficulty to all algebraically special vacuum fields.

If we take, e.g., a II-conformspinor, we may start with the "algebraic normal form" (71). Since $\varkappa^A$ determines a geodesic null congruence, we may again replace $\varkappa^A$ by $\alpha\varkappa^A$ and take the new $\varkappa^A$ to be constant along the rays, $\dot{\varkappa}_A = 0$. By a null rotation $\mu^A \longrightarrow \mu^A + B\,\varkappa^A$ we may introduce a new μ_A with $\dot{\mu}_A = 0$. Then the "algebraic normal form" goes over into the "analytic normal form"

$$\mathrm{II} = \lambda\,(\mathring{\mathrm{D}} - 3\,f\,\mathring{\mathrm{III}}) + \nu\,\mathring{\mathrm{N}} \tag{124}$$

where $\mathring{\mathrm{D}}$, $\mathring{\mathrm{III}}$, $\mathring{\mathrm{N}}$ denote conformspinors of the appropriate type which are constant along the rays. The advantage of this transformation is, of course, that the dependence of II_{ABCD} on v along a ray is completely given by the complex scalar factors λ, f, ϑ. Using the Bianchi identity and some of the propagation equations we can obtain propagation equations which can easily be integrated with the result (if $\theta \neq 0$)

$$II = z\overset{\circ}{N} + z^2\overset{\circ}{III} + z^3\overset{\circ}{II} + z^4\overset{\circ}{III}{}' + z^5\overset{\circ}{N}{}' \quad . \tag{125}$$

The index o indicates constancy along the rays. Again (123) is valid, the different terms have, therefore, a (luminosity-) distance dependence of the form $r^{-1}, \ldots, r^{-5}$, repsectively, and undergo duality rotations with angular velocities $\omega, \ldots, 5\omega$ where ω is the twist angular velocity of the congruence.

Obviously this decomposition of the curvature tensor is "intrinsic", i.e. independent of coordinate system, tetrad and parameterisation of the congruence.

This result (125) has been generalized by R. K. Sachs to vacuum fields of type I which contain a geodesic congruence the tangent of which is a principal null direction of the conformtensor. In this case the first three terms in (125) remain unchanged whereas the fourth term has to be replaced by $z^4\overset{\circ}{I}$; moreover, the power series in z does not break off as in the algebraically special cases but is, in general, infinite.

The results described so far are the main results of the propagation theory of "pure" gravitational waves apart from the important investigations concerning the explicit determination of the metrics of these fields. They support the conjectures which we have gained within the linear approximation and they give, even if it is granted that these "pure" fields are only approximate models for actual gravitational waves, valid asymptotically at large distances from the sources, a good insight into the kinematics of a gravitational wave. Moreover, they served as a starting point for the more involved investigations which we can only indicate in the last section.

9. Remarks on further investigations concerning gravitational waves.

A. Rigorous solutions

The explicit determination of vacuum metrics describing gravitational waves requires the calculation of the three Pfaffians $\Theta^{A\dot{B}}$; the fundamental form is then

$$G = -2(\Theta^{1\dot{1}}\Theta^{2\dot{2}} - \Theta^{1\dot{2}}\overline{\Theta^{1\dot{2}}}) .$$

The exterior differentials of these forms can be expressed, according to (89) and (99), as

$$(126) \qquad d\Theta^{A\dot{B}} = \Gamma^{A}{}_{CD\dot{E}}\,\Theta^{D\dot{E}}\wedge\Theta^{C\dot{B}} + \overline{\Gamma}^{\dot{B}}{}_{\dot{C}D\dot{E}}\,\Theta^{D\dot{E}}\wedge\Theta^{A\dot{C}}$$

by means of the spin coefficients and the forms themselves. The vacuum field equations are equivalent to the propagation equations for the spin coefficients described at the end of section 4, if the quantities R and S_{ABCD} are replaced by 0 .

If fields with an algebraically special conform tensor are to bi constructed, the spin basis $\{\varkappa, \mu\}$ and the $\Theta^{A\dot{B}}$ can be adapted to the congruence as described in section 5. Then some of the spin coefficients vanish (section 6) or can be made to vanish by transformations (58), (59). Accordingly, the equations (126) are simplified. The procedure for choosing preferred coordinates, cultivated mainly by Robinson and Trautman, can essentially be described as a repeated application of the theorem of Frobenius, combined with frame-transformations (58), (59). This theorem states: r Pfaffians θ^s on a V^n are locally linearly dependent on r functions x^s, $\vartheta^s = a^s{}_k dx^k$ (s, k = 1, ... r) , if and only if $d\theta^s \wedge \theta^1 \wedge \ldots \wedge \theta^r = 0$.

If coordinates have been chosen the field equations can either be set up directly in terms of the holonomic g_{ab}'s , or the equations (126) , written explicitely in terms of the coefficients of the $\theta^{A\dot{B}}$'s , can be combined with the Bianchi-identities and the propagation equations to yield a system of 34 differential equations for the components of the $\theta^{A\dot{B}}$'s , the spin coefficients, and the Γ_{ABCD} . This system of equations can then be solved successively by laborious calculations, we refer to the papers of Newman, Unti, Tamburino, and Kerr .

Especially the analogoues of th plane waves have been determined and extensively discussed. Locally, they behave precisely as was to be expected from the linear approximation. Globally, however, it seems to be impossible to satisfy the weak field conditions of section I. An important global property of the plane waves shall be mentioned : Their space-times are affinely complete, i.e. all their geodesics have infinite affine lengths in both directions. This property can perhaps be interpreted by considering these waves as completely "free" waves ; all the other known exact wave solutions contain singularities which have to be interpreted as some kind of sources.

B. Radiation from bounded sources, asymptotic behaviour.

The determination of the radiation field of a bounded source has been attacked mainly by Bondi, Sachs and collaborators. Since it is hopeless to determine the interior and exterior metric precisely, the approach of these authors has been to assume that space time is, outside of a time like cylinder, free of matter ($T_{ab}=0$) , topologically like a corresponding region in Minkowskian space time, and asymptotically flat in a certain sense. A coordinate system consisting of retarded time u, a luminosity distance r along the rays $k^a = u'^{,a}$, and two angles θ, φ is assumed such that the g_{ab} are analytic in $\frac{1}{r}$ for fixed u, θ, φ and approach the corresponding Minkowskian values if $r \rightarrow \infty$. The field equations are then used to obtain information about the leading terms of the g_{ab}'s and the R_{abcd}'s at large distances. A decomposition of the kind (125) is again obtained. Moreover, it is shown that if the field is static except in a certain retarded time interval (wave pulse) the mass of the source is smaller afterwards by an amount determined by the far field amplitudes (of the N, or $\frac{1}{r}$, part of $R...$) during the radiation interval. Moreover, these authors, in particular R.K. Sachs, have investigated the group of coordinate transformations preserving their boundary condition. These important investigations confirm the picture to which we have been led by the linear approximation and the "pure" radiation fields. The problem of existence of such Bondi-Sachs space times, however, and the interaction with the source are still open (and very difficult) questions.

We mention only that Penrose has reformulated the boundary condition by introducing ideal, infinite elements of the space time manifold.

C. Experimental verification

So far, no experimental or observational proof for the existence of gravitational radiation has been given. This is not surprising if we remember that, e.g., the gravitational radiation power emitted by the solar system, estimated with the linear approximation, is smaller by a factor 10^{24} than the emitted photon radiation.

The most serious effort to detect gravitational waves has been made by J. Weber. His apparatus consists of a 1,5 tons aluminium cylinder, the support of which is constructed such that mechanical disturbances (vibrations) are reduced as far as possible. Gravitational waves would excite normal vibrations of this cylinder, and corresponding piezoelectric voltages, highly amplified, could possibly be used to measure the absorbed radiation. So far, however, no positive result has been reported. Some methods to produce gravitational waves have also been discussed.

On the theoretical side, L. Halpern has calculated, with ordinary quantum number mechanics and the linear approximation of the gravitational theory, various transition probabilities for graviton-transitions in nuclei, atoms and molecules in order to find out whether their are cases which lead to measurable effects. Due to the smallness of $\frac{Gm^2}{e^2}$ the prospect of finding such cases is not large. For numerical values we refer to the cited paper.

References and Footnotes

1. See, e.g., R.H.Dicke, Sci.Am., 205,No.6, p.84 (1961), or B. Bertotti, D. Brill, R.Krotkov, ch.1 in Gravitation: An Introduction to Current Research, Ed. L.Witten, New York 1962.

2. Since eq. (3) of the text is not an ordinary conservation law it is necessary to have a physically reasonable argument for its validity which is independent of the field equation (1), if (3) is to be used as a restriction on admissable field equations. Such an argument has been given by J.L. Synge: If matter is treated as a gas the particles of which move along geodesics of the "macroscopic" metric g_{ab} except during collisions at which fourmomentum is conserved then the statistical energy-momentum tensor satisfies (3). See J.L.Synge, Relativity: The General Theory, Amsterdam 1960, pp. 165 - 167.

3. A more geometrical formulation of the contents of eq. (1) has been pointed out by J.A.Wheeler: Let Σ be an arbitrary space like hypersurface, u^a its unit normal ($u_a u^a = -1$), $d_{\mu\nu}$ its second fundamental tensor, $g_{\mu\nu}$ its induced metric, $\bar{R}$ the scalar curvature determined by $g_{\mu\nu}$, and $\mu = T_{ab}u^a u^b$ the energy density with respect to Σ. Then

$$\frac{1}{2}((d^\nu_\nu)^2 - d^\mu_\nu d^\nu_\mu - \bar{R}) = \mu, \qquad (g^{\nu\mu} d_{\mu\lambda} = d^\nu_\lambda \quad \text{etc.})$$

if postulated for arbitrary Σ's, is equivalent to (1).
See J.A.Wheeler, ch.4 in Gravitation and Relativity, ed by H.-Y. Chiu and W.F.Hoffmann, New York 1964.

4. To obtain equations of motion of test particles we follow the method of A. Papapetrou (Proc. Roy. Soc. 209, 248 (1951)) as modified by B. and

W. Tulczyjew (Recent Developments in General Relativity, New York 1962, p.465). Instead of working with a "dipole particle", however, we prefer to use the conditions $u^{[a}t^{b]c}u_c = 0$, $t^{abc}u_b u_c = 0$ on Tulczyjew's covariant moments t^{ab}, t^{abc} as definitions of the center of mass world line with four velocity u^a, and then evaluate the equations for the moments of orders zero and one approximately by neglecting those terms which are under the assumptions stated in the text numerically small in comparison with the leading terms. The resulting equations (4) do not differ sensibly from those derived for a pole-dipole particle by Papapetrou and by the Tulczyjews.

5a. In this connection it seems worthwhile to point out that the curvature tensor is a "state variable" of the gravitational field, i.e it is "algebraically" determined by the initial data on a space like hypersurface. In fact, the first and second fundamental forms of a hypersurface and their derivatives within this hypersurface determine, according to the generalized Gauss-Codazzi equations, some of the components of the space-time curvature tensor. The remaining components can be calculated algebraically from these and the energy momentum tensor if the symmetry properties of R.... and the field equations are used. The corresponding formulae can be said to express the "field-gradient" R_{abcd} in terms of "canonical" variables.

5 The set of all time-like geodesics determines the projective structure and, since it fixes the null-cone as boundary of the set of time like geodesics passing through one event, the conformal structure, too. These data determine the metric up to a constant factor according to H. Weyl, Mathematische Analyse des Raumproblems, Berlin 1923, 1. und 3. Vorlesung. Test particle experiments which can, at least in principle, be

used to measure space time intervals without use of standard clocks or rods have been devised independently by W. Kundt and B. Hoffmann (Recent Developments ..., see ref. 4, p. 303) and by R. Marzke (Princeton Senior thesis, 1959). The simple interpretation of the curvature tensor stated in the text in connection with eq. (5) is due to F.A.E. Pirani (Acta Phys. Polon., **15**, 389 (1956)).

6. We think of a macroscopic test particle such as an artificial satellite.

7. R.K. Sachs, Z. Physik **157**, 462 (1960), Appendix.

8. Put $F_{ab} = A_{ab} P(u)$ and assume that A_{ab} varies slowly in comparison with the phase factor P(u). Maxwell's equations then give (approximately) $F_{[ab} u,_{c]} = F^{ab} u,_{b} = 0$, i.e. F_{ab} is (nearly) a null field with propagation vector $u,^{a}$. Null fields propagate along null geodesics. Papers on electromagnetic null fields which contain also earlier results of Bel and Lichnerowicz are:
J. Robinson, Journ. Math. Phys., **2**, 290 (1961)
J. Ehlers and R.K. Sachs, Akad. Wiss. Lit. Mainz Abh. Math.-Nat. Kl. Jahrg. 1961, Nr. 1, Kap. 3, 3.

9. The linear approximation, its difficulties and generalizations can be used to motivate general relativity and to connect it with special relativistic field theories. Relevant papers are briefly reviewed and references are given by L. Halpern, Bull. Acad. Roy. Belgique ser. 5, **49**, 225 (1963).

10. See, e.g., V. Fock, The Theory of Space, Time, and Gravitation, London 1959.

11. A. Trautman, Bull. Acad. Polon. Sci. **6** (1958). The final paper was not available to me; in the preprint the theorem was stated without proof in a form slightly stronger than stated here, namely with 0 in place of $\varepsilon (> 0)$. I do not know a proof of the stronger version.

12. Kirchhoff's formula and an application of it which is similar to the one given in the text can be found, e.g., in Fock's book (ref. 10), §92.

13. Though we are not concerned with the quantum theory of gravitation in these lectures, I should like to describe the construction of the Hilbert space of "physical" one-graviton states because this deepens the analogy with electrodynamics and also adds to the understanding of the classical waves like (21). Consider the set E" of (real) solutions of (20) which admit a Fourier transform,

$$\psi_{ab}(x) = \int_C \tilde{\psi}_{ab}(k) e^{ik.x} dK \qquad (C: k^2 = 0), \qquad (A.1)$$

$$\tilde{\psi}_{ab} k^b = 0, \qquad (A.2)$$

$$\bar{\tilde{\psi}}_{ab}(-k) = \tilde{\psi}_{ab}(k). \qquad (A.3)$$

The restriction $\overset{+}{\tilde{\psi}}_{ab}$ of $\tilde{\psi}_{ab}$ to $C^{\uparrow}$ ($k^4 \geq 0$) therefore determines (and is determined by) ψ_{ab} uniquely. A gage-transformation which preserves (20) and whose generator ξ^a admits a Fourier transform $\tilde{\xi}^a$ induces the transformation

$$\overset{+}{\tilde{\psi}}_{ab} \longrightarrow \overset{+}{\tilde{\psi}}_{ab} + i(\Delta_{ab} \overset{+}{\tilde{\xi}}{}^c k_c - 2 \overset{+}{\tilde{\xi}}_{(a} k_{b)}). \qquad (A.4)$$

Consequently it is no loss of generality to impose the additional gage-condition

$$\overset{+}{\tilde{\psi}}{}^a{}_a = 0, \text{ or } \psi^a{}_a = 0; \qquad (A.5)$$

let E' be the corresponding subset of E". The remaining gage transformations are given by

$$\overset{+}{\tilde{\psi}}_{ab} \longrightarrow \overset{+}{\tilde{\psi}}_{ab} - 2i \overset{+}{\tilde{\xi}}_{(a} k_{b)}, \quad \overset{+}{\tilde{\xi}}{}^a k_a = 0. \qquad (A.6)$$

the integral

$$\langle \psi_{ab}, \varphi_{ab} \rangle \equiv \int_{C^{\uparrow}} \bar{\tilde{\psi}}{}^{ab} \tilde{\varphi}_{ab} dK \qquad (A.7)$$

defines a Lorents-invariant Hermitean form on the subset E of E' for which the (because of (A. 2) and (A. 5) non negative) functional $\langle \psi_{ab}, \psi_{ab} \rangle$ is finite. Because of the correspondence $\psi_{ab} \leftrightarrow \tilde{\psi}^{+}_{ab}$, E is in a natural way a complex vector space. The subspace N of E where $\langle \psi_{ab}, \psi_{ab} \rangle = 0$ consists precisely of the fields which are gage-equivalent to 0; the (completed) quotient space E/N is the required Hilbert space H of one graviton states. Eqs. (A. 1), (A. 3), and (A. 7) show: Orthochronous (antichronous) Lorentz-transformations induce in H unitary (antiunitary) mappings. The semi-unitary representation of the (inhomogeneous) Lorentz group thus constructed is irreducible; it belongs to mass 0 and spin 2 in the Wigner classification as can be seen from the invariants of the representation of the Lie algebra.

If we introduce along C (in k-space) a quasi-orthonormal tetrad $\{t, \bar{t}, k, m\}$ as defined by eqs. $(50)_1$ the general element of E can be represented as

$$\tilde{\psi}^{+}_{ab} = \underset{+}{A}\, t_a t_b + \underset{-}{A}\, \bar{t}_a \bar{t}_b + k_{(a} p_{b)}, \quad k^a p_a = 0, \qquad (A.8)$$

with two complex functions $\underset{+}{A}, \underset{-}{A}$ defined on $C^{\uparrow}$. Because of (A. 6) the last term can be gaged away. From (A. 7) we obtain $\langle \psi_{ab}, \psi_{ab} \rangle = \int_c (|\underset{+}{A}|^2 + |\underset{-}{A}|^2)\, dK$. H therefore consists of two orthogonal subspaces H_+, H_- which are both invariant under proper transformations, but are interchanged under improper transformations (such as P and T.) H_+ contains the states with positive helicity, H_- those with negative helicity.

Plane waves may be considered as improper elements of H. A monochromatic plane wave with positive (negative) helicity propagating in the z-direction is given, according to (A. 1) and (A. 8), by

$$\psi_{ab}(x) = \frac{A}{\sqrt{2}} \cdot \mathcal{R}\, (\overset{(-)}{t}_a \overset{(-)}{t}_b e^{i\omega(z-t)}) \quad A \text{ real}) \qquad (A.9)$$

which leads to a metric of the form (21), namely,

$$G = \underset{0}{G} + A((dx^2 - dy^2) \cos \omega(z-t) \underset{(+)}{-} 2\, dxdy \sin\omega(z-t)). \qquad (A.10)$$

In analogy with electrodynamics such a wave should be called left (right) circularly polarized. The intuitive interpretation of these as well as linear polarisation states in terms of test body behaviour follows from the discussion on pages 13 and 14 and the formula

$$R^{ab}{}_{cd} = 2\psi^{[a}{}_{[c|}{}^{|b]}{}_{d]} = -\sqrt{2}\, A\,\Re\left(\overset{(-)}{t}{}^{[a}{}_{k}{}^{b]}\,\overset{(-)}{t}{}_{[c}{}^{k}{}_{d]}\, e^{i\omega(z-t)}\right) \quad (A.11)$$

for the linearised curvature tensor associated with (A.9).

(The preceding considerations are formally similar to those in part VI of Lichnerowicz, Annli di Matematica ser. 4, 50, 1 (1960), though Lichnerowicz is concerned neither with the explicit construcion of the Hilbert space nor with that of the field operators but with the deduction of the form of the commutatiorrules. The actual construction of the Fock space and the field operator-distributions can be performed by using Wightman's improved version of the Gupta-Bleuler formalism.)

14. This metric results by superposing monochromatic plane waves (A.10) of both helicities travelling in the same direction.

15. The normal forms (22), (23) have been established by Bel, Lichnerowicz, and Pirani in their papers which marked the biginning of pure radiation theory; see the references in Pirani's article in the book "Gravitation" cited in 1. - Test body motion in exact plane waves has been studied in H. Bondi, F.A.E. Pirani, J. Robinson, Proc. Roy. Soc. A 251, 519 (1959) and

J. Ehlers and W. Kundt, ch. 2 in Gravitation, see ref. 1.

16. The definition of polarisation states by means of the curvature tensor has been given in 13. in connection with (A.11). That definition can be used, mutatis mutandis, in the full theory if R_{abcd} (s) is Fourier-analysed along

the world line of an observer.

17. This approach to gravitational radiation theory has been initiated by F.A. E. Pirani (Phys. Rev. 105, 1089 (1957)) and A. Lichnerowicz (C.R. Acad. Sc. Paris 246, 893 (1958)).

18. A. Trautman, see 11. - The expression (31) does not have the form $(23)_3$, but introducing a quasiorthonormal tetrad as defined in eqs. $(50)_1$ and taking into account $(30)_3$ and the vacuum field equation we may write

$$i_{ab} = \delta^c_a \, i_{cd} \, \delta^d_b = (k^c m_a + m^c k_a + t^c \bar{t}_a + \bar{t}^c t_a) \, i_{cd}(\ldots)$$

$$= 0(r^{-2}) + k_{(a} r_{b)} + r^{-1} 2 \Re (\ddot{N}_{cd} \bar{t}^c \bar{t}^d t_a t_b).$$ The $\frac{1}{r}$ - curvature tensor may therefore be expressed as $(23)_2$ with

$i_{ab} = \frac{2}{r} \Re (\ddot{N}_{cd} \bar{t}^c \bar{t}^d t_a t_b)$ which has the required form. The complex function $\ddot{N}_{ab} t^a \bar{t}^b$ of u, ω fully determines the far-field curvature tensor.

19. See R.K. Sachs, Proc. Roy. Soc. A 264, 309 (1961), sec. 2 and appendix B.

20. A careful discussion of the energy problem is given by A. Trautman, ch. 5 in the book "Relativity" cited in 1.

21. J. Weber, General Relativity and Gravitational Waves, Intersc., New York 1961, sec. 7.3.
J. Boardman and P. G. Bergmann, Phys. rev. 115, 1318 (1959) R. K. Sachs and P. G. Bergmann, Phys. Rev. 112, 674 (1958)

22. References can be found in Pirani's article mentioned in 15. and in R. K. Sach!s contribution to "Relativity, Groups, And Topology," Gordon and Breech, New York 1964.

23. The decomposition (35) is due to J. Geheniau and R. Debever (Bull. Acad. Roy. Belg., Cl. des Sc., 42 (1956).
The classification of conformtensors has been carried out by A. S. Petrov (Sci. Not. Kazan State University 114, 55 (1954)) and independently by J. Geheniau (C. R. Paris 244, 723 (1957)). The finer classification presented here has been given by L. Bel (These, Paris 1960). Our presentation follows R. Penrose (Ann. Phys. 10, 171 (1960)). More details are given in P. Jordan, J. Ehlers, R. K. Sachs (Akad. Wiss. Mainz 1, 1 (1961)) For a corresponding classification of "Ricci" tensors see R. V. Churchill, Trans. Am. Math. Soc. 34, 784 (1932).

24. See, e.g., W. L. Bade and H. Jehle, Rev. Mod. Phys. 25, 714 (1953). Some conventions used here differ from those in that paper. - For spinors in General Relativity, see Penrose, loc. cit. 23, and Jordan, Ehlers, Sachs loc. cit. 23.

25. H. Weyl, The Theory of Groups and Quantum Mechanics, London 1931 (reprinted by Dover Publ.), ch. III, sec. 8.
J. L. Synge, Relativity: The Special Theory, Amsterdam 1956, ch. IV, sects. 11, 12.

26. This method of decomposition is taken from a preliminary draft of a comprehensive book on spinors which is being written by R. Penrose and W. Rindler.

27. Cf. L. Witten, Phys. Rev. 113, 357 (1959),
R. Penrose loc. cit 23.

28. L. Bel (degenerate cases, cf. loc. cit. 23,
R. Debever, Bull. Soc. Math. Belgique 10 (1959) (general case)
R. Penrose (loc. cit. 23; spinor treatment).

29. P. Jordan, J. Ehlers, R. K. Sachs, loc. cit. 23.

30. K. Bichteler, Cartanformalisus für Spinoren, Preprint, Hamburg 1962; Z. Physik 178, 488 (1964). A similar calculus, based on the isomorphism between $\mathcal{L}_+^\uparrow$ and the complex orthogonal group in three dimensions in place of (53), has been developed by R. Debever, M. Cahen and L. Defrise; cf. R. Debever, Le Rayonnement Gravitationnel, preprint, Brussels 1964, chap. III.

31. E. Newman and R; Penrose, Journ. Math. Physics 3, 566 (1962)

32. R. K. Sachs, Proc. Roy. Soc. A 264, 309 (1961), sec. 4; P. Jordan, J. Ehlers, R. K. Sachs, loc. cit. 23.

33. Cf. references 31. and 32.

34. This theorem emerged out of the work of L. Mariot, A. Lichnerowicz, L. Bel, J. Robinson, A. Trautman, R. Debever, R. K. Sachs; for references, see R. K. Sachs, loc. cit. 32.

35. J. N. Goldberg and R. K. Sachs, Acta Phys. Polon. 22, 13, (1962), E. Newman and R. Penrose, loc. cit. 31.

36. A conformally invariant generalisation of the Goldberg-Sachs theorem has been proven by W. Kundt and A. Thompson (C. R. 254, No. 25 (1962)) and by J. Robinson and A. Schild (Journ. Math. Phys. 4, 484 (1963)).

37. This and other "distances" are discussed by R. K. Sachs in *Recent developments in general relativity*, Warsaw 1962, p. 395

38. These theorems on the propagation of algebraically special vasuum curvature tensors have been obtained by R. K. Sachs; full proofs are given in P. Jordan, J. Ehlers, R. K. Sachs, loc. cit. 23. For the special case of twist-free

congruences somewhat stronger statements have been obtained by J. Robinson and A. Trautman
(Proc. Roy. Soc. A 265, 463 (1962)).

39. R. K. Sachs, loc. cit. 32, sec. 6.

40. J. Robinson and A. Trautman, loc. cit 38. See also W. Kundt, Z. Physik 163, 77 (1961), W. Kundt, and M. Trümper Akad. Wiss. Mainz Nr. 12, 1962.

41. Cf., e.g., A. Lichnerowicz, Theorie globale des Connexions et des groupes d'Holonomie, Paris 1955.
The observation that this theorem is a natural tool for introducing preferred coordinates has been pointed out by K. Bichteler, cf. the references 30.

42. E. Newman and T. Unti, J. Math. Phys. 3, 896 (1962), E. Newman and L. A. Tamburino, J. Math. Phys. 2, 667 (1961), R. P. Kerr, Phys. Rev Letters 11, 237 (1963).

43. Cf. the last two of the references 15.

44. H. Bondi, M. van der Burg, A. Metzner, Proc. Roy. Soc. A 269, 21 (1962) R. K. Sachs, Proc. Roy. Soc. A 270, 103 (1962)

45. R. K. Sachs, Asymptotic Symmetries in Gravitational Theory, USASRDL, 1962.

46. R. Penrose, Phys. Rev. Letters 10, 66 (1963),
R. Penrose, Conformal Treatment of Infinity, contribution to Relativity, Groups, and Topology cited in 22.

47. Cf., e.g., ref 10, p. 404

48. See, e.g. J. Weber, General Relativity and Gravitational Waves, New York 1961, and the contribution of this author to "Relativity, Groups, and

Topology" cited in 22.

49. L. Halpern and B. Laurent, Nuovo Cimento, 1964.

CENTRO INTERNAZIONALE MATEMATICO ESTIVO

(C. I. M. E.)

L. BEL

SUR TROIS PROBLEMES PHYSIQUES RELATIFS AU ds^2 DE SCHWARZSCHILD

SUR TROIS PROBLEMES PHYSIQUES RELATIFS AU ds^2 DE SCHWARZSCHILD

par L. BEL

INTRODUCTION GENERALE

Ce cours est consacré à l'étude de trois problèmes relatifs au contenu physique du ds^2 de Schwarzschild.

Ces trois problèmes sont successivement traités aux chapitres II, III et IV pour les quels nous avons choisi respectivement les titres suivants :

Quantification des états liés gravitationnels.

Ondes planes à l'infini.

Deux aspects de la généralisation du groupe de Lorentz inhomogène.

Les résultats du premier chapitre concernant le probème de la quantification des états liés gravitationnels résultent de l'application au ds^2 de Schwarzschild de deux méthodes de quantification différentes. La première est la méthode de Bohr pour la quantification des orbites circulaires. Elle confirme les résultats que l'on obtient dans l'étude des états liés des particules d'épreuve soumises à un potentiel central newtonnien. La deuxième méthode est la méthode du type mécanique ondulatoire. Elle confirmerait les résultats précédents sauf qu'elle prévoit l'existence d'une état supplémentaire, qu'on peut appeler super-lié, pour lequel l'énergie de liaison est d'un ordre de grandeur bien plus important que celle qui correspond aux états du spectre prévu par la méthode de Bohr.

Le chapitre III vise à la généralisation de la notion d'onde plane, familière en Relativité Restreinte. A cette notion se rattachent au moins deux problèmes physiques. Le problème de la diffusion des particules d'épreuve dans un champ de Schwarzchild, des deux points de vue: classique et quantique. Et le problème de la caractérisation du rayonnement émis par objects à distance infinie, d'òu l'utilité des résultats obtenus pour l'étude des phénomènes de l'aberration et l'effet Döppler.

Nous obtenons au chapitre IV des formules de transformation qui généralisent les formules de transformation qui définissent le groupe de Lorentz inhomogène. Plus précisément, ces formules de transformation sont fonction de la masse centrale qui crée le champ de Schwarzschild et se réduisent aux formules de Lorentz si on fait tendre la masse centrale vers zéro. L'interprétation cinématique est conservée par cette généralisation.

Le chapitre I est un chapitre de Rappels, auquel nous aurions pu inclure les trois appendices qui terminent ces Notes. Les rappels du chapitre I ont un caractère trés général et peuvent être utiles pour la lecture des trois chapitres qui suivent, tandis que les trois appendices se rapportent uniquement à la section II du chapitre II. Ces rappels, aussi que les appendices, ne prétendent pas être un résumé équilibré des questions traitées car nous avons mis l'accent sur les aspects auxquels nous devions par la suite nous référer.

Nous avons rédigé pour chaque chapitre des introductions plus détaillées auxquelles peut utilement se rapporter le lecteur qui désirerait avoir une vue d'ensemble plus complète de ces Notes avant d'en enterprendre la lecture.

Je remercie très sincèrement le Centro Internazionale Matematico Estivo et particulièrement M. le Professeur C. Cattaneo de m'avoir invité à faire un cours sur ces sujets à la Sauze d'Oulx en été 1964.

L. Bel

CHAPITRE I : RAPPELS

SECTION I : QUELQUES RAPPELS DE MECANIQUE ANALYTIQUE

1. - Définitions générales et principaux résultats.

a. - Du point de vue de la mécanique analytique un système mécanique holonome à n degrés de liberté peut être considéré défini par sa fonction de Lagrange.

$$L(q^i, \dot{q}^i, t)$$

qui est une fonction des n variables q^i choisis pour décrire la configuration du système, des dérivées $\dot{q}^i$ de ces variables par rapport au temps t, et dans le cas général du temps lui-même.

On dit que l'espace des variables (q^i) est l'espace de configuration, et on peut aussi dire que l'espace des variables (q^i, t) est l'espace de configuration des évènements.

Les moments conjugués des variables q^i sont par définition :

$$p_i \equiv \frac{\partial L}{\partial \dot{q}^i} \qquad i = 1, 2, \ldots, n$$

Ces n définitions peuvent être considérées comme un système d'équations sur les $\dot{q}^i$. Le système mécanique est régulier, ce qui est supposé ici, si ce système d'équations peut être résolu. Dans ce cas nous pouvons obtenir les fonctions

$$\dot{q}^j = \dot{q}^j (q^i, p_i, t) \tag{I, 1}$$

L'action du système le long d'un chemin donné d'équations $q^i = q^i(t)$, origine $q^i_o = q^i(t_o)$ et d'extrémité $q^i_1 = q^i(t_1)$, est par définition

$$S = \int_{t_o}^{t_1} L(q^i, \dot{q}^i, t)dt \; . \tag{I, 2}$$

l'intégrale étant calculée le long du chemin envisagé.

Les équations du mouvement du système sont les équations d'Euler des extrémales de l'intégrale ci-dessus, c'est-à-dire :

(I, 3) $$\frac{d}{dt}p_i - \frac{\partial L}{\partial q^i} = 0 \qquad i = 1, 2, \ldots, n$$

Une fonction $P(q^i, \dot{q}^i, t)$ est dite être une intégrale première des équations précédentes si

$$\frac{dP}{dt} = \frac{\partial P}{\partial t} + \frac{\partial P}{\partial q^i}\dot{q}^i + \frac{\partial P}{\partial \dot{q}^i}\ddot{q}^i = 0$$

Il est clair que si $\frac{\partial L}{\partial q^j} = 0$, c'est-à-dire si la variable q^j est cyclique, le moment conjugué p_j est une intégrale première.

b. - Il existe trois méthodes pour obtenir les solutions du système de n équations du deuxième ordre (I, 3) correspondant à des conditions initiales données : $q^i(t_o) = q^i_o$; $\dot{q}^i(t_o) = \dot{q}^i_o$.

La première consiste évidemment dans l'approche direct du système d'équations précédent et ne fait intervenir de ce fait aucune autre grandeur nouvelle.

La deuxième méthode consiste à ramener les équations (I, 3) à un système de 2n équations du premier ordre, ce qui peut se faire de plusieurs manières. La méthode des équations canoniques de Hamilton en est une. En voici les deux étapes principales.

Le Hamiltonien du système est par définition :

$$H = p_i\dot{q}^i - L$$

C'est une fonction des $q^i, \dot{q}^i$ et t mais qui compte tenu de (I, 1) peut être considéré aussi comme une fonction des q^i, p_i et t, et être ainsi désigné par :

$$H(q^i, p_i, t)$$

Si le lagrangien ne depend pas explicitement du temps, il est une intégrale première de (I, 3).

Les équations canoniques de Hamilton sont le système de 2n équations:

$$\dot{q}^i = \frac{\partial H}{\partial p_i} \quad , \quad \dot{p}_i = -\frac{\partial H}{\partial q^i} \; ; \tag{I, 4}$$

système d'équations qui est équivalent à (I, 3) dans ce sens que si $q^i(t)$ et $p_i(t)$ est la solution de (I, 4) correspondant aux conditions initiales $q^i_o = q^i(t_o)$ et $p_{i_o} = p_i(t_o)$, $q^i(t)$ est la solution de (I, 3) correspondant aux conditions initiales $q^i_o = q^i(t_o)$ $\dot{q}^i_o = \dot{q}^i(q^j_o, p_{i_o}, t_o)$ et réciproquement.

Enfin la troisième méthode, celle qu'il nous intéresse particulièrement de rappeler, est la méthode de l'intégrale complète de l'équation de Hamilton-Jacobi.

L'équation de Hamilton-Jacobi est l'équation en derivées partielles du premier ordre

$$\frac{\partial S}{\partial t} = - H\left(q^i, \frac{\partial S}{\partial q^i}, t\right) \tag{I, 5}$$

obtenue en remplaçant dans l'Hamiltonien p_i par $\frac{\partial S}{\partial q_i}$ et égalant l'expression obtenue à $-\frac{\partial S}{\partial t}$.

Suivant la terminologie de la théorie des équations en dérivées partielles du premier ordre on appelle intégrale complète une solution de (I, 5) contenant n+1 constantes d'intégration indépendantes. Il est clair que la fonction S n'apparaissant dans (I, 5) que par ces dérivées partielles, une de ces constantes est toujours une constante additive.

Soit $S(q^i, t/P_i)$ une telle intégrale complète, où P_i sont les n constantes d'intégration autres que la constante additive. Les n constantes P_i sont indépendantes si

$$\text{dét}\left|\frac{\partial^2 S}{\partial q^j \partial P_k}\right| \neq 0 \tag{I, 6}$$

On peut donc résoudre le système d'équations sur les P_i

$$p_j = \frac{\partial S}{\partial q^j}(q^i, t \; P_i)$$

Les fonctions :

$$P_i = P_i(q^j, p_j, t)$$

ainsi obtenues sont, comme on peut le voir, des intégrales premières du système d'équations (I, 3).

Posons

$$S(q^i, t \,|\, P_i \,|\, q^i_o, t_o) \equiv S(q^i, t \,|\, P_i) - S(q^i_o, t_o \,|\, P_i) \quad .$$

Pour obtenir la solution de (I, 3) correspondant aux conditions initiales $q^i_o = q^i(t_o)$, $p_{i_o} = p_i(t_o) = p_i(q^j_o, \dot{q}^j_o, t_o)$ à partir de l'intégrale complète $S(q^i, t \,|\, P_i)$, il suffit de résoudre le système d'équations sur les q^i :

$$\frac{\partial S}{\partial P_i}(q^j, t \,|\, P_i(q^k_o, p_{k_o}, t_o) \,|\, q^j_o, t_o) = 0$$

qui est encore résoluble d'après (I, 6). Les fonctions

$$q^i = q^i\left[t;\, q^j_o, t_o;\, p_{j_o}(q^k_o, \dot{q}^k_o, t_o)\right]$$

donnent la solution désirée.

Pour des valeurs des P_i fixes les équations :

$$P_i = P_i\left[q^i, p_i(q^j, \dot{q}^j, t), t\right] \tag{I, 7}$$

définissent un système d'équations du premier ordre. Les fonctions P_i étant des intégrales premières du système (I, 3), les solutions de (I, 7) sont des solutions de (I, 3). Nous appellerons l'ensemble de ces solutions le faisceau associé aux P_i fixes choisis.

c. - Rappelons aussi la définition de variété transversale.

Toute extrémale de (I, 2) peut être représentée dans l'espace de configuration des évènements par un arc de courbe d'équations

$$q^i = q^i(t) \qquad t = t$$

En un point q_1^i, t_1 de cette extrémale nous pouvons calculer par conséquent H et p_i.

Définition. - Une surface de l'espace de configuration d'équation locale $S(q^i, t) = cte.$, est dite être une variété transversale à une extrémale de (I, 2) si au point (q_1^i, t_1) d'intersection de l'extrémale avec la surface on a :

$$\frac{\partial S}{\partial t} = -H \qquad \frac{\partial S}{q^i} = p_i$$

Avec cette définition on peut énoncer le résultat suivant :

Si $S(q^i, t / P_i)$ est une intégrale complète de (I, 5), chacune des surfaces $S(q^i, t \mid P_i) = cte$ pour P_i fixes est une variété transversale à chacune des extrémales des faisceaux associé aux P_i.

Enfin, le dernier résultat qu'il nous convient de rappeler est le théorème suivant :

Théorème. - Si $S(q^i, t / P_i)$ est une intégrale complète de (I, 5) et (q_o^i, t_o), (q_1^i, t_1) sont les deux points extrêmes d'un arc d'extrèmale du faisceau associé aux P_i, on a :

$$S(q_1^i, t_1 \mid P_i \mid q_o^i, t_o) = \int_{t_o}^{t_1} L(q^i, \dot{q}^i, t)dt$$

l'intégrale étant calculée le long de l'extrémale envisagée. Autrement dit $S(q_1^i, t_1 \mid P_i \mid q_o^i, t_o)$ est l'action du système le long de l'arc d'extrémale considérée.

SECTION II : LE ds^2 DE SCHWARZSCHILD

2. - Le système mécanique associé au ds^2 de Schwarzschild.

a. - Le ds^2 de Schwarzschild est le ds^2 décrivant le champ de gravitation d'une masse à symétrie sphérique. Il est de ce fait la généralisation relativiste du potentiel newtonnien de cette masse.

Nous écrivons le ds^2 de Schwarzschild sous sa forme habituelle en coordonnées polaires

$$ds^2 = \sigma\, dt^2 - \frac{1}{\sigma c^2}\, dr^2 - \frac{r^2}{c^2}\,(d\theta^2 + \sin^2\theta\; d\varphi^2) \tag{I, 8}$$

avec

$$\sigma \equiv 1 - \frac{2\mu}{r} \qquad \mu \equiv \frac{kM}{c^2} \tag{1, 9}$$

k étant la constante de la gravitation, M la masse du corps qui crée le champ et c la vitesse de la lumière dans le vide.

Quant ceci sera plus commode nous poserons

$$\xi^o = t\,,\ \xi^1 = r\,,\ \xi^2 = \theta\,,\ \xi^3 = \varphi$$

et :

$$ds^2 = g_{\alpha\beta}\, d\xi^\alpha\, d\xi^\beta \qquad \alpha, \beta = 0, 1, 2, 3$$

Les potentiels sont donc :

$$g_{oo} = \sigma \quad g_{11} = -\frac{1}{\sigma c^2}\,,\ g_{22} = -\frac{r^2}{c^2}\,,\ g_{33} = -\frac{r^2\sin^2\theta}{c^2} \qquad g_{\alpha\beta} = 0 \ \text{si}\ \alpha \neq \beta \tag{I, 10}$$

de même

$$g^{oo} = \frac{1}{\sigma} \quad g^{11} = -\sigma c^2\,,\ g^{22} = -\frac{c^2}{r^2}\,,\ g^{33} = -\frac{c^2}{r^2\sin^2\theta} \qquad g^{\alpha\beta} = 0 \ \text{si}\ \alpha \neq \beta \tag{I, 11}$$

et

$$g = \text{dét}\,\left|g_{\alpha\beta}\right| = -\frac{r^4\sin^2}{c^6} \tag{I, 12}$$

b. - Au ds^2 ci-dessus correspond le système mécanique défini par le Lagrangien :

$$L = -mc^2 \Sigma$$

où m est la masse de la particule d'épreuve soumise au champs de gravitation de M et où :

$$\Sigma \equiv \sqrt{\sigma - \frac{\dot{r}^2}{c^2} - \frac{r^2}{c^2}(\dot{\theta}^2 + \sin^2\theta\, \dot{\varphi}^2)} \tag{I, 13}$$

Les trois moments conjugués sont :

$$p_r = \frac{\partial L}{\partial \dot{r}} = \frac{m\dot{r}}{\sigma \Sigma}, \quad p_\theta = \frac{mr^2}{\Sigma}, \quad p_\varphi = \frac{mr^2 \sin^2\theta\, \dot{\varphi}}{\Sigma} \tag{I, 14}$$

Le système mécanique est régulier car les équations précédentes peuvent être résolues pour $\dot{r}$, $\dot{\theta}$ et $\dot{\varphi}$. En effet nous pouvons écrire tout d'abord :

$$\dot{r} = \frac{\sigma \Sigma}{m} p_r, \quad \dot{\theta} = \frac{\Sigma}{mr^2} p_\theta, \quad \dot{\varphi} = \frac{\Sigma}{mr^2 \sin^2\theta} p_\varphi ; \tag{I, 15}$$

et utilisant ces expressions dans la définition de Σ (I, 13) il vient

$$\Sigma = \sqrt{\sigma - \frac{\sigma \Sigma^2}{c^2 m^2} p_r^2 - \frac{\Sigma^2}{c^2 r^2 m^2}\left(p_\theta^2 + \frac{1}{\sin^2\theta} p_\varphi^2\right)}$$

d'où on tire l'expression de Σ en fonction des p :

$$\Sigma = \frac{\sqrt{\sigma}}{\sqrt{1 + \frac{1}{c^2 m^2}\left[\sigma\, p_r^2 + \frac{1}{r^2}\left(p_\theta^2 + \frac{1}{\sin^2\theta} p_\varphi^2\right)\right]}}$$

Avec cette expression pour Σ les formules (I, 15) donnent le résultat désiré.

L'action de la particule de masse m le long d'un chemin dont les points extrêmes correspondent aux valeurs t_o et t_1 est

$$S = \int_{t_o}^{t_1} - mc^2 \Sigma \, dt \tag{I, 16}$$

l'intégrale étant calculée le long du chemin envisagé.

c. - Les équations du mouvement de m sont les équations d'Euler des extrémales de l'intégrale ci-dessus qui correspondent à Σ réel positif. Elles s'écrivent compte tenu de (I, 13) et (I, 14) :

$$\frac{d}{dt}\left(\frac{m\dot{r}}{\sigma\Sigma}\right)+\frac{mc^2}{2\Sigma}\left[\left(1+\frac{\dot{r}^2}{c^2\sigma^2}\right)\frac{d\sigma}{dr}-\frac{2r}{c^2}\left(\dot{\theta}^2+\sin^2\theta\,\dot{\varphi}^2\right)\right]=0$$

(I, 17)
$$\frac{d}{dt}\left(\frac{mr^2\dot{\theta}}{\Sigma}\right)-\frac{mr^2\sin\theta\cos\theta\,\dot{\varphi}^2}{\Sigma}=0$$

$$\frac{d}{dt}\left(\frac{mr^2\sin^2\theta\,\dot{\varphi}}{\Sigma}\right)=0$$

La dernière de cas équations exprime évidemment que l'expression entre parenthèses est une intégrale première. Nous poserons

(I, 18)
$$M\equiv\frac{mr^2\sin^2\theta\,\dot{\varphi}}{\Sigma}$$

Il est clair que par un choix convenable de l'axe polaire nous pouvons faire de sorte que $\dot{\varphi}_o=\dot{\varphi}(t_o)=0$ et par conséquence $M=0$. L'équation (I, 18) nous dit dans ces circonstances que $\dot{\varphi}=0$. Autrement dit que φ = cte et que par conséquent les trajectoires des particules d'épreuve sont contenues dans un plan.

d. - Des trois méthodes rappelées dans la section précédente pour résoudre les équations (I, 17), nous allons appliquer celle de l'intégrale complète de l'équation de Hamilton-Jacobi.

Construisons tout d'abord l'Hamiltonien correspondant. De (I, 14) il vient:

$$H=\frac{m}{\Sigma}\left[\frac{\dot{r}^2}{\sigma}+r^2(\dot{\theta}^2+\sin^2\theta\,\dot{\varphi}^2)\right]+mc^2\Sigma$$

et de (I, 13) :

(I, 19)
$$H=\frac{mc^2\sigma}{\Sigma}$$

H est une intégrale première puisque L ne dépend pas explicitement du temps.

C'est l'intégrale de l'énergie. Nous poserons

$$E = H$$

On remarquera que H, et donc E, est toujours positif.

D'après l'expression de $\sum$ en termes des p, l'équation de Hamilton-Jacobi est

$$\frac{\partial S}{\partial t} = -mc^2\sqrt{\sigma}\sqrt{1 + \frac{1}{c^2m^2}\left\{\sigma\left(\frac{\partial S}{\partial r}\right)^2 + \frac{1}{r^2}\left[\left(\frac{\partial S}{\partial \theta}\right)^2 + \frac{1}{\sin^2\theta}\left(\frac{\partial S}{\partial \varphi}\right)^2\right]\right\}}$$

Après élévation au carré cette équation peut s'écrire :

$$\text{(I, 20)} \qquad \Delta_1 S \equiv \frac{1}{\sigma}\left(\frac{\partial S}{\partial t}\right)^2 - c^2\sigma\left(\frac{\partial S}{\partial r}\right)^2 - \frac{c^2}{r^2}\left[\left(\frac{\partial S}{\partial \theta}\right)^2 + \frac{1}{\sin^2\theta}\left(\frac{\partial S}{\partial \varphi}\right)^2\right] = m^2c^4$$

e. - On peut trouver une intégrale complète de cette équation par la méthode habituelle. Prenons comme fonction d'essai une fonction S de la forme:

$$S = -Et + W(r) + Q(\theta) + M\varphi$$

où E et M sont deux constantes. La substitution de cette fonction d'essai dans l'équation (I, 20) donne :

$$\text{(I, 21)} \qquad \frac{E^2}{\sigma^2} - c^2\sigma\left(\frac{dW}{dr}\right)^2 - \frac{c^2}{r^2}\left[\left(\frac{dQ}{d\theta}\right)^2 + \frac{M^2}{\sin^2\theta}\right] = m^2c^4$$

Posons encore

$$\left(\frac{dQ}{d\theta}\right)^2 + \frac{M^2}{\sin^2\theta} = P^2$$

où P est une nouvelle constante. L'équation (I, 21) devient :

$$\frac{E^2}{\sigma} - c^2\sigma\left(\frac{dW}{dr}\right)^2 - \frac{c^2}{r^2}P^2 = m^2c^4$$

De ces deux dernières équations on tire

$$\frac{dW}{dr} = \frac{\varepsilon}{c\sigma}\,\Omega \qquad\qquad \varepsilon^2 = +1$$

où

(I, 22)
$$\Omega \equiv \sqrt{E^2 - \frac{c^2P^2}{r^2}\sigma - m^2c^4\sigma}$$

et :

$$\frac{dQ}{d\theta} = \gamma\sqrt{P^2 - \frac{M^2}{\sin^2\theta}} \qquad \gamma^2 = +1$$

Nous avons ainsi obtenu l'intégrale complète que voici

(I, 23)
$$S = -Et + \frac{\mathcal{E}}{c}\int\frac{\Omega}{\sigma}dr + \gamma\int\sqrt{P^2 - \frac{M^2}{\sin^2\theta}}\,d\theta + M\varphi$$

f. - Pour obtenir l'interprétation des constantes M et P nous procédons comme il a été indiqué dans la section précédente. Il vient tout d'abord d'après (I, 14) :

$$\frac{mr^2\sin^2\theta\,\dot{\varphi}}{\Sigma} = p_\varphi = \frac{\partial S}{\partial\varphi} = M$$

M est donc l'intégrale première (I, 18) déjà trouvée.

De la définition de P et de l'expression ci-dessus de M il vient ensuite :

$$P^2 = \left(\frac{\partial S}{\partial\theta}\right)^2 + \frac{M^2}{\sin^2\theta} = p_\theta^2 + \frac{M^2}{\sin^2\theta} = \frac{m^2r^4}{\Sigma^2}(\dot{\theta}^2 + \sin^2\theta\,\dot{\varphi}^2)$$

Enfin pour E , puisque

$$\frac{\partial S}{\partial t} = -E$$

E n'est autre que l'Hamiltonien du système, c'est-à-dire l'énergie de la particule.

D'autre part de

$$p_r = \frac{\partial S}{\partial r} = \frac{dW}{dr} = \frac{\mathcal{E}\Omega}{c\sigma}, \quad p_\theta = \frac{\partial S}{\partial\theta} = \frac{dQ}{d\theta} = \gamma\sqrt{P^2 - \frac{M^2}{\sin^2\theta}}$$

et de (I, 15), il vient

$$\dot{r} = \frac{\varepsilon \Sigma \Omega}{cm} \qquad \dot{\theta} = \frac{\gamma \Sigma}{mr^2} \sqrt{P^2 - \frac{M^2}{\sin^2\theta}}$$

ε et γ sont donc respectivement $\frac{\dot{r}}{|\dot{r}|}$ et $\frac{\dot{\theta}}{|\dot{\theta}|}$.

g. - Nous allons poursuivre la résolution du problème de l'intégration du système d'équations (I, 17) nous bornant aux particules pour lesquelles $M = 0$ ce qui revient comme nous l'avons vu à un choix particulier de l'axe polaire.

Avec les notations de la section précédente la fonction $S(\xi^i, t/E, P/\xi^i_o, t_o)$ est dans ces conditions :

$$S(\xi^i, t/E, P/\xi^i_o, t_o) = -E(t-t_o) + \frac{\varepsilon}{c}\int_{r_o}^{r} \frac{\Omega}{\sigma}\, dr + P(\theta - \theta_o), \quad P = \gamma\sqrt{P^2} \tag{I, 24}$$

et les équations du mouvement sont par conséquent d'après (I, 22) et (I, 24) :

$$\begin{aligned} \frac{\partial S}{\partial E} &= -(t-t_o) + \frac{\varepsilon}{c}\int_{r_o}^{r} \frac{E}{\sigma\Omega}\, dr = 0 \\ \frac{\partial S}{\partial P} &= -\varepsilon c P\int_{r_o}^{r} \frac{dr}{r^2\Omega} + \theta - \theta_o = 0 \end{aligned} \tag{I, 25}$$

La première de ces équations permet d'obtenir la fonction

$$r = r(t;\, r_o, t_o, E, P)\ ;$$

la deuxième est l'équation de la trajectoire dans le plan φ = cte qui dépend évidemment des conditions initiales.

Nous terminerons ces rappels par deux remarques.

Remarque I . Il est clair par inspection de (I, 11) et (I, 20) que cette dernière n'est autre que l'équation

$$\Delta_1 S \equiv g^{\alpha\beta} \frac{\partial S}{\partial \xi^\alpha} \frac{\partial S}{\partial \xi^\beta} = m^2 c^4\ ;$$

c'est pour cette raison que nous avions déjà utilisé la notation Δ_1 qui est la notation habituelle pour le premier paramètre différentiel de Beltrami. Ceci n'est pas particulier au ds^2 de Schwarzschild. On peut voir, en effet, par une méthode intrinsèque directe que ce résultat subsiste pour tout Lagrangien de la forme $L = -mc^2 \frac{ds}{dt}$, ds^2 étant une métrique riemannienne. Nous avons employé les méthodes classiques de la mécanique analytique plutôt que des méthodes géométriques intrinsèques parce qu'elles s'adaptent mieux aux interprétations physiques.

Remarque II . Soient $\dot{r}, \dot{\theta}, \dot{\varphi}$ les composantes de la vitesse en un point (r, θ, φ, t) d'une particule décrivant une extrémale G de (I, 16). Soit $S(r, \theta, \varphi, t)$ = cte une surface de l'espace de configuration des évènements coupant l'extrèmale G au point considéré. D'après la définition de la section précédente S = cte est une variété transversale à l'extrèmale G si au point d'intersection :

$$\frac{\partial S}{\partial t} = -M \; ; \; \frac{\partial S}{\partial r} = p_r \; , \; \frac{\partial S}{\partial \theta} = p_\theta \; , \; \frac{\partial S}{\partial \varphi} = p_\varphi \qquad \text{(I, 26)}$$

En termes de géométrie riemannienne G est une géodésique de (I, 8). Le vecteur tangent, normé à mc^2, à cette géodésique est le vecteur de composantes contravariantes

$$u^o = \frac{mc^2}{\Sigma} \; , \quad u^1 = \frac{mc^2\dot{r}}{\Sigma} \; , \quad u^2 = \frac{mc^2\dot{\theta}}{\Sigma} \; , \quad u^3 = \frac{mc^2\dot{\varphi}}{\Sigma}$$

les composantes covariantes étant

$$u_o = \frac{mc^2\sigma}{\Sigma} \quad u_1 = -\frac{m\dot{r}}{\sigma\Sigma} \quad u_2 = -\frac{mr^2\dot{\theta}}{\Sigma} \quad u_3 = -\frac{mr^2\sin^2\theta\,\dot{\varphi}}{\Sigma}$$

c'est-à-dire, d'après (I, 14) et (I, 19)

$$u_o = M \; , \quad u_1 = -p_r \; , \quad u_2 = -p_\theta \; , \quad u_3 = -p_\varphi$$

On peut donc traduire les formules (I, 26) par :

$$\frac{\partial S}{\partial \xi^\alpha_M} = -u_\alpha$$

et dire que la surface $S = cte$ est une variété transversale à la géodésique G si elle est orthogonale a cette géodésique au sens de la métrique envisagée. Ce résultat est aussi valable pour tout Lagrangien de la forme $L = -mc^2 \frac{ds}{dt}$.

CHAPITRE II : QUANTIFICATION DES ETATS LIES

Introduction. -

Nous abordons ici le problème de la quantification des états liés d'origine purement gravitationnel.

L'energie potentielle d'une masse m dans le champ de gravitation newtonnien d'une masse M ne diffère de l'énergie potentielle d'une charge -e dans le champ de Coulomb d'une charge e que par les valeurs numériques de ces quantités et des constantes caractéristiques. Les mêmes règles de quantification peuvent donc être appliquées pour déterminer les états liés de deux masses on de deux charges de signes opposés. On obtient ainsi les formules suivants pour les énergies des états liés de deux masses

$$W_n = -\lambda^2 \frac{mc^2}{2n^2} \qquad n = 1, 2, \dots$$

où

$$\lambda \equiv \frac{k\,Mm}{\hbar c}$$

Cette constante λ est la constante qui au point de vue quantique joue le rôle qui correspond à $\gamma = \frac{e^2}{\hbar c}$ dans le cas de l'interaction de deux charges +e et -e.

Pour des masses M et m de l'ordre de la masse du nucleon, disons, λ est de l'ordre de 10^{-40} et par conséquent l'énergie de liaison correspondante est toujours tout à fait négligéable. Ce simple résultat numérique expliquerait pourquoi les effets gravitationnels newtonniens seraient indetéctables dans le domaine microscopique.

L'object du présent chapitre est d'examiner le même problème à la lumière des méthodes de quantification plus ou moins raffinées qu'on peut appliquer à la théorie du ds^2 de Schwarzschild.

La première de ces méthodes sera la méthode de Bohr pour la quantification des orbites circulaires. Cette méthode confirme qualitativement, et même quantitativement à un très haut dégré d'approximation, les résultats issus de la théorie de Newton et rappelés ci-dessus.

La deuxième méthode est la méthode du type mécanique ondulatoire. Nous avons choisi comme fonction d'onde l'équation

$$(\Delta_2 + \frac{m^2c^4}{\hbar^2})\Psi = 0$$

Δ_2 étant le laplacien intrinsèque sur les fonctions pour le ds^2 de Schwarzschild. L'étude des solutions correspondant aux fonction d'état des états liés, avec un principe de quantification approprié, conduit à ranger ceux-ci en deux groupes. Le premier groupe, qui comporte un nombre infini dénombrable d'états, correspond au spectre obtenu précédemment et, d'être seul, confirmerait les résultats connus. Le deuxième groupe comporte un seul état dont l'énergie de liaison est de l'ordre de $0,3mc^2$ et serait donc très importante, même pour des masses de l'ordre de grandeur envisagé ci-dessus.

SECTION I : QUANTIFICATION DES ORBITES CIRCULAIRES

1. - Les orbites circulaires.

a. - Considérons une extrémale de (I, 16) dont la trajectoire soit un cercle de rayon $a > 0$. On sait que ceci existe et nous nous proposons de caractériser ces solutions particulières des équations du mouvement par une relation entre les deux constantes du mouvement E et P . Nous supposons, bien entendu, ce cercle contenu dans un plan φ = cte.

Posons :

$$u = \frac{1}{r}$$

Si $r = a$, $u = \frac{1}{a} \equiv \alpha$ et par conséquent toutes les dérivées de u par rapport à θ doivent être nulles. Or de (I, 25) il vient

$$\frac{du}{d\theta} = -\frac{\varepsilon\Omega}{cP} \tag{II, 1}$$

Ω^2 étant d'après (I, 22) et (I, 9) :

$$\Omega^2 = 2\mu\chi^2 u^3 - \chi^2 u^2 + 2\mu\, m^2 c^4 u - \xi^2 \tag{II, 2}$$

où nous avons posé :

$$\chi = cP \qquad \xi^2 = m^2 c^4 - E^2$$

Nous devons par conséquent avoir

$$\Omega^2(\alpha) = 0 . \tag{II, 3}$$

D'autre part par définition de (II, 1) il vient

$$\frac{d^2u}{d\theta^2} = -\frac{\varepsilon}{2cP\Omega}\frac{d}{du}\Omega^2\frac{du}{d\theta}$$

Soit :

$$\frac{d^2u}{d\theta^2} = \frac{1}{2c^2P^2}\,\frac{d}{du}\,\Omega^2$$

et cette expression dévant être nulle pour $u = \alpha$

(II, 4) $$\frac{d}{du}\,\Omega^2(\alpha) = 0$$

Reciproquement il est facile de voir que (II, 3) et (II, 4) entraînent $u = \alpha$. Elles caractérisent donc les orbites circulaires.

Les conditions (II, 3) et (II, 4) expriment comme on sait que $u = \alpha$ est une racine double de l'équation :

$$\Omega^2(u) = 0\ .$$

b. - Nous obtiendrons par conséquent la caractérisation désirée des orbites circulaires si nous obtenons la relation entre E et P qui fait que le polynome Ω^2 est de la forme :

$$\Omega^2(u) = 2\mu\chi^2(u-\alpha)^2(u-\beta)$$

β étant la troisième racine. Le développement de l'expression ci-dessus est:

$$\Omega^2(u) = 2\mu\chi^2u^3 - 2\mu\chi^2(2\alpha+\beta)u^2 + 2\mu\chi^2(\alpha^2+2\alpha\beta)u - 2\mu\chi^2\alpha^2\beta$$

De l'identification de l'expression précédente à (II, 2) il vient :

(II, 5) $$2\mu(2\alpha+\beta) = 1,\quad \chi^2(\alpha^2+2\alpha\beta) = m^2c^4,\quad 2\mu\chi^2\alpha^2\beta = \xi^2$$

Il s'agit maintenant d'éliminer α et β entre ces trois relations pour obtenir la relation entre ξ et χ et par conséquent entre E et P. Nous obtenons tout d'abord

$$\beta = \frac{1}{2\mu} - 2\alpha\ ;$$

puis, l'équation pour α :

$$-3\alpha^2 + \frac{\alpha}{\mu} - \frac{m^2c^4}{\chi^2} = 0$$

dont les deux solutions sont

$$\alpha = \frac{1}{6\mu}\left(1 + \varepsilon\sqrt{1 - \frac{12m^2c^4\mu^2}{\chi^2}}\right) \qquad \varepsilon = \pm 1$$

Les valeurs de α et β ainsi obtenues substituées dans la dernière des relations (II, 5) nous donnent la formule désirée :

$$\xi^2 = \frac{\chi^2}{2.\,3^3\mu^2}\left[-1 + \frac{3^2.\,2m^2c^4\mu^2}{\chi^2} - \varepsilon\left(1 - \frac{12m^2c^4\mu^2}{\chi^2}\right)^{3/2}\right]$$

2. - Quantification.

a. - Introduisons maintenant le postulat de qualification. Nous poserons d'après la méthode de Bohr

$$P = n\hbar$$

et conviendrons que les seules valeurs possibles pour P sont ceux qui correspondent à n entier positif. $h = 2\pi\hbar$ est la constante de Planck.

Compte tenu du fait que :

$$\frac{1}{\mu} = \frac{c^2}{kM} = \frac{1}{\lambda}\frac{mc}{\hbar}, \qquad \frac{m^2c^4\mu^2}{\chi^2} = \frac{\lambda^2}{n^2}$$

où λ est la constante sans dimensions :

$$\frac{kMm}{\hbar c}$$

nous pouvons écrire :

(II, 6) $$\alpha = \frac{mc}{6\hbar\lambda}\left(1 + \varepsilon\sqrt{1 - \frac{12\lambda^2}{n^2}}\right)$$

et l'expression correspondante pour ξ^2 est :

$$\xi^2 = \frac{n^2m^2c^4}{2.\,3^3\lambda^2}\left[-1 + \frac{3^3.\,2\lambda^2}{n^2} - \varepsilon\left(1 - \frac{12\lambda^2}{n^2}\right)^{3/2}\right]$$

b. - A chaque valeur de n , la formule précédente fait correspondre une valeur E_n de l'énergie dont nous allons obtenir une expression plus simple en développant le terme entre crochets par rapport au paramètre λ que nous supposerons beaucoup plus petit que 1 . De l'ordre de 10^{-40} si l'on veut.

De :

$$(1 - \frac{12\lambda^2}{n^2})^{3/2} = 1 - \frac{3^2 . 2}{n^2}\lambda^2 + \frac{3^3 . 2}{n^4}\lambda^4 + 0(\lambda^6)$$

il vient :

$$\xi^2 = \frac{n^2 m^2 c^4}{2 . 3^3 \lambda^2}\left[-1 + \frac{3^3 . 2\lambda^2}{n^2} - \mathcal{E}(1 - \frac{3^2 . 2}{n^2}\lambda^2 + \frac{3^3 . 2}{n^4}\lambda^4 + 0(\lambda^6))\right.$$

Pour $\mathcal{E} = -1$ nous obtenons :

$$\xi^2 = \frac{m^2 c^4}{n^4}\lambda^2 + 0(\lambda^4) \; ; \qquad \text{(II, 7)}$$

et pour $\mathcal{E} = +1$:

$$\xi^2 = - \frac{n^2 m^2 c^4}{3^3 \lambda^2} + \frac{2}{3} m^2 c^4 - \frac{m^2 c^4}{n^2}\lambda^2 + 0(\lambda^4) \qquad \text{(II, 8)}$$

c. - Si nous nous intéressons uniquement aux états liés au sens stricte, c'est-à-dire ceux pour lesquels on a :

$$\xi^2 = m^2 c^4 - E^2 > 0$$

il est clair que seul $\mathcal{E} = -1$ doit être retenu. En effet $\mathcal{E} = +1$ conduit à $\xi^2 < 0$. Bornons-nous pour l'instant aux états liés au sens stricte.

De (II, 7) il vient à l'approximation envisagé :

$$E_n = mc^2 \sqrt{1 - \frac{\lambda^2}{n^2}}$$

d'où finalement :

$$E_n = mc^2 - \lambda^2 \frac{mc^2}{2n^2} \qquad \text{(II, 9)}$$

Autrement dit les énergies de liaison des différents états sont :

$$-W_n = \lambda^2 \frac{mc^2}{2n^2}$$

et coïncident avec les énergies correspondantes du cas newtonnien.

Nous obtenons les rayons des orbites correspondant à chaque valeur de n a partir de (II, 6) en posant $\mathcal{E} = -1$. Le résultat à des termes de l'ordre de λ^3 près est :

$$\alpha_n = \frac{mc\lambda}{\hbar n^2} ;$$

donc :

$$a_n = \frac{1}{\alpha_n} = \frac{1}{\lambda} \frac{\hbar n^2}{mc}$$

résultat qui coincide encore avec le résultat newtonnien.

d. - Les résultats précédents peuvent être considérés satisfaisants dans ceci qu'ils confirment à un très haut degré d'approximation les résultats d'une théorie, la théorie de Newton, que l'on sait être une excellente théorie du champ de gravitation.

Toutefois nous croyons qu'il ne serait pas justifié de prétendre que la question des énergies de liaison des états liés gravitationnels est tranchée. Il convient de se rappeler que nous n'avons quantifié que les orbites circulaires et que la condition d'orbite circulaire élimine l'essentiel des corrections relativistes.

Même s'il faut se méfier de tirer des conclusions trop générales de l'étude que nous venons de faire, il nous semble qu'elle a été utile. Il a été en effet conjecturé [3] que dans le problème qui nous occupe il n'y aurait pas d'états liés quantifiés. Les résultats que nous avons exposés semblent prouver le contraire et on ne voit pas très bien pourquoi la méthode de quantification de Bohr, même en la sachant incomplète, serait inadéquate dans notre cas alors qu'elle a très bien réussi ailleurs.

e. - Nous avons exclu l'expression (II, 8) de ξ^2 qui correspondait à $\mathcal{E} = +1$ parce qu'elle entraînait $\xi^2 < 0$ et il ne s'agirait plus d'états liés au sens stricte. Il est clair pourtant que s'agissant toujours d'orbites circulaires, les états envisagés sont des états liés dans ce sens que la trajectoire n'a pas de branches à l'infini. Il s'agit donc d'états liés virtuels instables. Qualitativement ce résultat est curieux mais quantitativement il n'est pas intéressant. On peut voir, en effet, par (II, 8) que les énergies correspondantes sont de l'ordre de $\lambda^{-1} mc^2$ et par conséquent énormes pour des valeurs de λ physiquement raisonnables.

SECTION II : METHODE GENERALE

3. - L'équation d'onde.

Pour appliquer les méthodes de la mécanique ondulatoire au problème de la quantification des états liés gravitationnels le premier problème qui se pose est de choisir convenablement une équation d'onde.

Il est conforme à l'esprit de la Relativité Générale de considérer comme libre toute particule d'épreuve sur laquelle n'agit d'autre force que le champ de gravitation. Si en outre la particule d'épreuve que l'on considère à un spin zéro, l'équation d'onde qu'il convient de choisir doit être la généralisation naturelle de l'équation d'onde des particules libres et spin zéro de la Relativité Restreinte.

Cette généralisation naturelle est :

$$(\Delta_2 + \frac{m^2 c^4}{\hbar^2}) \Psi = 0 \qquad \text{(II, 10)}$$

Δ_2 étant le laplacien intrinsèque sur les fonctions pour le ds^2 envisagé. Plus précisément :

(II, 11) $$\Delta_2 \Psi = \frac{1}{\sqrt{|g|}} \frac{\partial}{\partial \xi^\alpha} \left(\sqrt{|g|}\, g^{\alpha\beta} \frac{\partial \Psi}{\partial \xi^\beta} \right)$$

$$\xi^0 = t, \ \xi^1 = \tau, \ \xi^2 = \theta, \ \xi^3 = \varphi$$

Si le ds^2 envisagé était le ds^2 de Minkowski l'équation (II, 10) serait l'équation de Klein-Gordon.

En portant dans (II, 11) les potentiels de gravitation (I, 11) correspondant au ds^2 de Schwarzschild on obtient, tous calculs faits, pour l'équation (II, 10) l'expression :

(II, 12) $$\left(\sigma^{-1} \frac{\partial^2}{\partial t^2} + c^2 P^2_{(r)} + \frac{c^2}{r^2} \Lambda^2_{(\theta,\varphi)} + \frac{m^2 c^4}{\hbar^2}\right) \Psi = 0$$

où

$$P^2_{(r)} \equiv - \left[\sigma \frac{\partial^2}{\partial r^2} + \left(\frac{2\sigma}{r} + \frac{p}{r^2}\right) \frac{\partial}{\partial r} \right] \qquad \sigma = 1 - \frac{2\mu}{r} \equiv 1 - \frac{p}{r}$$

et où $\Lambda^2_{(\theta,\varphi)}$ est, au facteur $\hbar^2$ près, l'opérateur carré du moment cinétique :

$$\Lambda^2_{(\theta,\varphi)} \equiv - \left[\frac{1}{\sin\theta} \frac{\partial}{\partial \theta} \left(\sin\theta \frac{\partial}{\partial \theta}\right) + \frac{1}{\sin^2\theta} \frac{\partial^2}{\partial \varphi^2} \right]$$

4. - Séparation des variables angulaires et temporelle. Equation radiale.

Nous allons nous intéresser aux solutions de l'équation (II, 10) qui sont en même temps fonctions propres des opérateurs énergie et carré du moment cinétique, c'est-à-dire solutions de :

$$i\hbar \frac{\partial \Psi}{\partial t} = E \Psi, \quad \hbar^2 \Lambda^2_{(\theta,\varphi)} = L^2 \Psi$$

E et L étant des constantes dont la première doit être identifiée à l'énergie de la particule d'épreuve. La première de ces équations entraîne que Ψ doit être de la forme :

$$\Psi = \chi(r, \theta, \varphi) e^{-i\frac{Et}{\hbar}}$$

Quant à la deuxième, si on impose en outre l'uniformité de la fonction Ψ par rapport aux variables θ et φ elle entraîne :

$$L^2 = \hbar^2 \, l(l+1) \qquad \text{avec l'entier} \geqslant 0$$

et

$$\chi = R(r) Y_l(\theta, \varphi)$$

Y_l étant une harmonique sphérique associée à la valeur de l choisie. La fonction Ψ doit donc être finalement de la forme :

(II, 13) $$\Psi = R(r) Y_l e^{-i\frac{E}{\hbar}t}$$

où il ne reste qu'à déterminer la fonction $R(r)$.

La substitution de l'expression (II, 13) dans (II, 12) donne, compte tenu des résultats précédents, l'équation radiale suivante pour R :

(II, 14) $$\frac{d^2R}{dr^2} + \frac{2r-p}{r(r-p)} \frac{dR}{dr} + \frac{-\kappa^2 r^3 + q r^2 - l(l+1)(r-p)}{r(r-p)^2} R = 0$$

où nous avons posé

(II, 15) $$\kappa^2 \equiv \frac{1}{c^2 \hbar^2} (m^2 c^4 - E^2) \qquad q \equiv \frac{m^2 c^2}{\hbar^2} p$$

Nous nous bornerons à considérer des états liés au sens stricte, c'est-à-dire ceux pour lesquels :

$$\kappa^2 > 0$$

et pour fixer les idées nous poserons :

$$\kappa = + \sqrt{\kappa^2}$$

5. - Forme limite de l'équation radiale pour $c \to \infty$.

Avant de procéder à l'étude systématique de l'équation (II, 14) nous nous proposons d'établir la forme limite de cette équation quand c tend vers infini. Le principe de ce calcul est basé sur l'identification de la constante E avec l'intégrale première de l'énergie (I, 19) des équations des extrémales de (I, 16) :

$$E = \frac{mc^2\sigma}{\sqrt{\sigma - \frac{\dot{r}^2}{\sigma c^2} - \frac{r^2}{c^2}(\dot{\theta}^2 + \sin^2\theta\,\dot{\varphi}^2)}}$$

Les comportements limites suivants sont évidents d'après les définitions correspondantes

(II, 16) $$\lim_{c\to\infty} p = 0 \qquad \lim_{c\to\infty} \sigma = 1$$

D'autre part q ne dépend pas en fait de c :

(II, 17) $$q = \frac{2kMm^2}{\hbar^2}$$

Il ne nous reste donc qu'à établir le comportement limite de $\mathcal{H}^2$. Or :

$$m^2c^2 - \frac{E^2}{c^2} = mc^2\left[1 - \frac{\sigma^2}{\sigma - \frac{\dot{r}^2}{\sigma c^2} - \frac{r^2}{c^2}(\dot{\theta}^2 + \sin^2\theta\,\dot{\varphi}^2)}\right]$$

ou encore :

$$m^2c^2 - \frac{E^2}{c^2} = mc^2\left[1 - \frac{\sigma^2}{\sigma - \frac{v^2}{c^2} + \frac{\dot{r}^2}{c^2}(1 - \frac{1}{\sigma})}\right]$$

où :

$$v^2 = \dot{r}^2 + r^2(\dot{\theta}^2 + \sin^2\theta\,\dot{\varphi}^2)$$

et d'après (I, 1) :

$$m^2c^2 - \frac{E^2}{c^2} = m^2c^2\left[1 - \frac{1 - \frac{4kM}{rc^2} + \frac{4k^2M^2}{r^2c^4}}{1 - \frac{1}{c^2}(\frac{2kM}{r} + v^2) + 0(\frac{1}{c^4})}\right]$$

Par conséquent :

$$\lim_{c\to\infty} m^2c^2 - \frac{E^2}{c^2} = \lim_{c\to\infty} m^2c^2 \left[1 - (1 - \frac{4kM}{rc^2} + 4\frac{k^2M^2}{r^2c^4})(1 + \frac{2kM}{rc^2} + \frac{v^2}{c^2} + 0(\frac{1}{c^4}))\right]$$

Donc :

$$\lim_{c\to\infty} \mathcal{H}^2 = -\frac{2m}{\hbar^2} W \tag{II, 18}$$

où :

$$W \equiv \frac{1}{2} mv^2 - \frac{kMm}{r}$$

serait l'énergie totale newtonnienne d'une particule de masse m dans le champ de gravitation de M.

En substituant (II, 16), (II, 17) et (II, 18) dans (II, 14) nous obtenons pour c tendant vers infini la forme limite :

$$\frac{d^2R}{dr^2} + \frac{2}{r}\frac{dR}{dr} + \frac{2m}{\hbar^2}\left[W + \frac{kMm}{r} - \frac{\hbar^2 l(l+1)}{2mr^2}\right] R = 0$$

qui n'est autre que la partie radiale de l'équation de Schrodinger pour le potentiel newtonnien (Voir App. B).

On sait que la théorie du ds^2 de Schwarzschild se réduit à la théorie de Newton du potentiel central quand c tend vers infini. D'après le résultat précédent nous voyons qu'il en va également de même pour l'aspect ondulatoire envisagé ici. Ce qui d'ailleurs ajoute une justification supplémentaire au fait d'avoir considéré l'équation (II, 10) comme fonction d'onde des particules d'épreuve de spin zéro dans un champ de Schwarzschild.

6. - Forme canonique de l'équation radiale. Potentiel effectif.

Toute équation différentielle linéaire du second ordre peut toujours être ramenée à una forme canonique dans laquelle la dérivée première de la fonction inconnue n'y figure pas. Dans le cas de l'équation (II, 14) nous arrivons à la forme canonique en posant (Voir App. A):

$$R = \frac{F}{\sqrt{r(r-p)}}$$

et nous obtenons pour F l'équation :

$$\frac{d^2F}{dr^2} + \left[-\mathcal{H}^2 + U(\mathcal{H}, 1, r) \right] F = 0$$

où :

$$\text{(II, 19)} \qquad U(\mathcal{H}, 1, r) \equiv \frac{(-2p\mathcal{H}^2+q)r^3 + \left[1(1+1)-p^2\mathcal{H}^2\right]r^2 + 1(1+1)pr + \frac{1}{4}p^2}{r^2(r-p)^2}$$

L'équation précédente se présente comme la partie radiale d'une équation de Schrödinger, disons, pour laquelle le potentiel effectif serait U . A propos de ce potentiel deux faits sont à signaler [Comparer avec (B, 4) et (C, 4)] :

a) Le potentiel effectif dépend, à part de 1 et de r , de $\mathcal{H}$ et donc de E

b) Le comportement à l'infini est en r^{-1} sauf si

$$-2p\mathcal{H}^2 + q = 0$$

auquel cas le comportement est en r^{-2}.

La valeur de E qui satisfait l'équation précédente est :

$$E = \frac{1}{\sqrt{2}} mc^2$$

Nous retrouverons plus tard cette valeur de E comme valeur de l'énergie possible d'un état super-lié. Cet état, s'il existait vraiment, n'aurait pas d'analogue dans la théorie quantique de l'interaction newtonnienne.

7. - Les points singuliers à distance finie. (1)

a.- Ecrivons l'équation (II, 14) sous la forme :

(II, 20) $$P_o \frac{d^2R}{dr^2} + P_1 \frac{dR}{dr} + P_2 R = 0$$

avec :

$$P_o \equiv r(r-p)^2 \qquad P_1 \equiv (2r-p)(r-p)$$

$$P_2 \equiv -\varkappa^2 r^3 + qr^2 - l(l+1)(r-p)$$

Les points singuliers à distance finie de l'équation (II, 20) c'est-à-dire les racines de $P_o = 0$, sont par conséquent

$$r = 0 \qquad \text{et} \qquad r = p$$

b. - Posons

$$Q_o = r^{-1} P_o \qquad Q_1 = P_1 \qquad Q_2 = rP_2$$

et écrivons les polynomes Q_i (i = 1, 2, 3) sous la forme :

$$Q_i = \sum_{h=0}^{4} Q_{i,h} r^h$$

Nous aurons

(II, 21) $$Q_{o,o} = p^2 \qquad Q_{1,o} = p^2 \qquad Q_{2,o} = 0$$

En particulier, puisque $Q_{o,o} \neq 0$ nous voyons que le point $r = 0$ est un point singulier régulier. L'équation indicielle

$$s(s-1)Q_{o,o} + sQ_{1,o} + Q_{2,o} = 0$$

en ce point est d'après (II, 21):

$$s^2 = 0$$

(1) Voir App. A.

Nous sommes dans le cas où l'équation indicielle a une racine double et par conséquent au voisinage de $r = 0$ il n'existe qu'une seule solution régulière au sens stricte, c'est-à-dire ne contenant pas de logaritmes. Cette solution régulière sera de la forme

(II, 22) $$R = \sum_{h=0}^{\infty} a_h r^h$$

a_o étant arbitraire mais différent de zéro et les autres coefficients devant être calculés par les formules (A, 7). La série (II, 22) sera convergente dans le cercle de centre $r = 0$ et rayon p.

c. - Les polynomes P_i peuvent être considérés comme des polynomes en (r-p), ce que nous indiquerons en les désignant dans ce cas par $\tilde{P}_i$.

Posons :

$$\tilde{Q}_o = (r-p)^{-2}\tilde{P}_o \qquad \tilde{Q}_1 = (r-p)^{-1}\tilde{P}_1 \qquad \tilde{Q}_2 = \tilde{P}_2$$

et

$$\tilde{Q}_i = \sum_{h=0}^{3} Q_{i,h}(r_1 - p)^h$$

Nous aurons :

$$\tilde{Q}_{o,o} = p \qquad \tilde{Q}_{1,o} = p \qquad \tilde{Q}_{2,o} = -\varkappa^2 p^3 + qp^2$$

Puisque $\tilde{Q}_{o,o} \neq 0$, le point $r = p$ est lui aussi un point singulier régulier. L'équation indicielle est :

$$s^2 - \varkappa^2 p^2 + qp = 0$$

ou encore, compte tenu des définitions (II, 15) :

$$s^2 = -\frac{E^2 p^2}{c^2 \hbar^2}$$

dont les deux racines sont :

$$s_1 = i\frac{Ep}{c\hbar} \qquad s_2 = -i\frac{Ep}{c\hbar}$$

$s_1 - s_2$ n'étant pas un entier, l'équation (II, 20) admet par conséquent au voisinage de $r = p$, un sistème fondamental de deux solutions régulières au sens stricte, de la forme [3]:

$$R = (r-p)^{\varepsilon \frac{iEp}{c\hbar}} \sum_{n=0}^{\infty} a_n (r-p)^n \qquad \varepsilon = \pm 1 \qquad a_o \neq 0$$

on si l'on veut de la forme :

$$R = e^{\varepsilon \frac{iEp}{c\hbar} \log(r-p)} \sum_{n=0}^{\infty} a_n (r-p)^n$$

Les coefficients a_n se calculent par les formules correspondantes à (A, 7). La série ainsi construite ne sera en général convergente que dans le cercle de centre $r = p$ et rayon p.

D'après les résultats des App. B et C on aurait pu espérer que ou bien le comportement de la solution régulière au point $r = 0$, ou bien au point $r = p$ dependrait de 1 . Les résultats ci-dessus prouvent qu'il n'en est rien.

8. - Solutions au voisinage de l'infini. (1)

a. - Revenons aux polynomes Q_i . Les dégrés de ces polynomes sont :

$$\text{dig } Q_o = 2 \qquad \text{dig } Q_1 = 2 \qquad \text{dig } Q_2 = 4$$

Donc :

$$Q_{o,4} = 0 \qquad Q_{1,4} = 0 \qquad Q_{2,4} \neq 0$$

et par conséquent l'équation indicielle à l'infini :

$$\rho(\rho-1)Q_{o,\nu} + \rho Q_{1,\nu} + Q_{2,\nu} = 0$$

où

$$\nu \equiv \max \text{dig } Q_i = 4$$

(1) Voir App. A.

n'a pas de solutions. Autrement dit, le point de l'infini est un point singulier irrégulier et par conséquent il n'existe pas de solutions régulières au voisinage de l'infini. Par contre il peut exister des solutions normales comme nous allons le voir (Comparer avec App. B et C).

b. - Pour r tendant vers l'infini l'équation (II, 14) devient :

$$\frac{d^2R}{dr^2} - \varkappa^2 R = 0$$

qui admet les solutions

$$R = Ae^{\varepsilon \varkappa r} \qquad A : \text{const} \qquad \varepsilon = \pm 1$$

$\varkappa$ étant positif ε doit être -1 si l'on veut que R soit nulle à l'infini.

Ceci suggère donc de chercher des solutions de la forme

(II, 23) $$R = V(r)e^{-\varkappa r}$$

Après ce changement de variable l'équation pour V est

(II, 24) $$\bar{P}_o \frac{d^2V}{dr^2} + \bar{P}_1 \frac{dV}{dr} + \bar{P}_2 V = 0$$

avec :

(II, 25)
$$\bar{P}_o \equiv r(r-p)^2 \qquad P_1 \equiv -2\varkappa r^3 + 2(1+2\varkappa p)r^2 - p(3+2\varkappa p)r + p^2$$
$$\bar{P}_2 \equiv (q-2\varkappa-2p\varkappa^2)r^2 + \left[p^2\varkappa^2 + 3p\varkappa - l(l+1)\right]r + p\left[l(l+1) - \varkappa p\right]$$

Posons :

(II, 26) $$\bar{Q}_o = r^{-1}\bar{P}_o \quad , \quad \bar{Q}_1 = \bar{P}_1 \quad , \quad \bar{Q}_2 = r\bar{P}_2$$

Nous avons maintenant

$$\bar{\nu} = \max \text{dig}\, Q_i = 3$$

Ecrivons les polynomes Q_i sous la forme :

$$\overline{Q}_i = \sum_0^3 \overline{Q}_{i,h} r^h$$

De (II, 25) et (II, 26) il vient :

$$\overline{Q}_{o,3} = 0 \qquad \overline{Q}_{1,3} = -2\mathcal{H} \qquad \overline{Q}_{2,3} = q - 2\mathcal{H} - 2p\mathcal{H}^2$$

$\overline{Q}_{o,3}$ étant nul et $\overline{Q}_{1,3}$ non nul, le point $r = \infty$ est un point quasi-régulier de l'équation (II, 24). L'équation indicielle est maintenant :

(II, 27) $$-2\mathcal{H}\rho + q - 2\mathcal{H} - 2p\mathcal{H}^2 = 0$$

dont la solution est :

$$\rho = \frac{q-2p\mathcal{H}^2}{2\mathcal{H}} - 1$$

c. - Supposons $\mathcal{H}$ connu. ρ est donc connu, et s'il existe une solution régulière au voisinage de l'infini, elle sera de la forme

(II, 28) $$V = r^{\rho} \sum_{m=0}^{\infty} b_m r^{-m} \qquad b_o \neq 0$$

Si cette solution existe, (II, 23) sera une solution normale de l'équation (II, 20). Les coefficients b_m doivent être calculés par les formules (A, 10) et la solution régulière V existera si ces coefficients conduisent à un développement convergent pour $r > p$. Autrement la formule (II, 28) manque de signification fonctionnelle.

Toutefois d'après un théorème de Poincaré [2] relatif à la méthode de Laplace pour la résolution d'équations différentielles linéaires dont les coefficients sont des polynomes, on peut affirmer qu'il existe toujours une solution au voisinage de l'infini qui coincide avec (II, 28) si cette dernière existe et dans le cas contraire a pour développement asymptotique le deuxième membre de (II, 28).

Nous ne savons pas à l'heure actuelle si le développement (II, 28) est convergent ou pas, mais donnons quand-même le relation de ricurrence pour le calcul des coefficients b_m puisque nous savons, d'après le théorème précédent, que de toutes façons ce développement a une signification.

La formule de récurrence pour $m \geqslant 0$ est :

$$b_{m+3} f_3(\rho, m) + b_{m+2} f_2(\rho, m) + b_{m+1} f_1(\rho, m) + b_m f_o(\rho, m) = 0$$

où :

$$f_3(\rho, m) \equiv 2\kappa m + q - 2\kappa(\rho - 2) - 2p\kappa^2$$

$$f_2(\rho, m) \equiv m^2 - (2\rho - 3 + 4\kappa p)m + (\rho - 2)(\rho - 1 + 4\kappa p) + p^2\kappa^2 + 3\kappa p - l(l+1)$$

$$f_1(\rho, m) \equiv -2pm^2 + p(4\rho - 3 + 2\kappa p)m - p(\rho - 1)(2\rho - 1 + 2\kappa p) + pl(l+1) - \kappa p^2$$

$$f_o(\rho, m) \equiv p^2 m^2 - 2p^2\rho\, m + p^2\rho^2$$

Elle doit être complétée par les deux relations suivantes qui donnent b_1 et b_2, b_o étant arbitraire :

$$b_1\left[-2\kappa(\rho - 1) + q - 2\kappa - 2p\kappa^2\right] + b_o\left[\rho(\rho - 1) + 2\rho(1 + 2\kappa p) + p^2\kappa^2 + 3p\kappa - l(l+1)\right] = 0$$

$$b_2\left[-2\kappa(\rho - 2) + q - 2\kappa - 2p\kappa^2\right] + b_1\left[(\rho - 1)(\rho - 2) + 2(\rho - 1)(1 + 2\kappa p + p^2\kappa^2 + 3p\kappa - l(l+1)\right] +$$
$$+ b_o\left[-2p\rho(\rho - 1) - p\rho(3 + 2\kappa p) + l(l+1) - p^2\kappa\right] = 0$$

9.- Le postulat de quantification.

a. - Si dans l'équation indicielle (II, 27) :

$$2p\kappa^2 + 2\kappa(\rho + 1) - q = 0 \qquad \text{(II, 29)}$$

nous supposons ρ connu, elle devient une équation en κ et par conséquent en E.

Dans les cas considérés aux App. B et C ρ est déterminé à un entier $n' \geqslant 0$ près par les conditions imposées à la fonction R d'être normale à l'infini et régulière au sens stricte à l'origine. Le raccordement de la solution normale à l'infini à la solution régulière à l'origine ne présente pas de difficulté.

Ici le problème de raccordement de la solution au voisinage de l'infini aux solutions régulières aux points $r = 0$ et $r = p$ se présente sous une forme autrement plus compliquée et nous ne savons pas à l'heure actuelle comment ce problème doit être résolu. Nous sommes donc obligés, si nous voulons poursuivre cet étude des états liés, de postuler les valeurs possibles de ρ .

Nous postulons que ρ doit être entier.

Il y a deux justifications initiales à ce postulat.

a) Il est exactement satisfait dans le problème considéré à l'Appendice B.

b) Il est très approximativement satisfait dans le problème considéré à l'Appendice C . En effet, dans ce cas, les valeurs possibles de ρ ne diffèrent de valeurs entières que par des termes qui sont au maximum de l'ordre de $\gamma^2 \simeq (\frac{1}{137})^2$. Donc tres petits. Or il suffit de comparer (C, 13) à (II, 9) ou (B, 12) pour se rendu compte que dans le problème qui nous occupe c'est la constante λ qui joue le rôle de la constante de structure finie γ . Pour des valeurs de λ suffisamment petites, on peut donc espérer que le postulat énoncé se trouvera être très proche du résultat correct.

Il existe une troisième justification de laquelle nous parlerons au paragraphe suivant.

b. - De l'équation (II, 29) il vient :

$$\mathcal{H} = -\frac{\rho+1}{2p} + \sqrt{\left(\frac{\rho+1}{2p}\right)^2 + \frac{a}{2p}} \tag{II, 30}$$

Nous n'avons retenu que le signe + de la racine carrée car autrement $\mathcal{H}$ serait négatif contrairement à ce que nous avons supposé.

De :

$$\mathcal{H} \equiv \frac{1}{c\hbar} \sqrt{m^2c^4 - E^2} \leq \frac{mc}{\hbar}$$

il resulte :

$$\rho \geqslant -1-\lambda \qquad \text{où} \qquad \lambda \equiv \frac{kMm}{\hbar c} < 1$$

Les valeurs possibles de ρ compatibles avec le postulat de quantification et avec la définition de $\mathcal{V}$ sont donc finalement :

$$\rho : -1, 0, 1, 2, \ldots$$

10. - Les états newtonniens.

La troisième justification du postulat de quantification adopté nous la tirons de la valeur de l'énergie qui correspond aux valeurs de ρ autres que -1 . Energie qui à la constante additive mc^2 près, comme il se doit, n'est autre que l'énergie des niveaux newtonniens, ou, autrement dit, l'énergie des états liés de l'équation (B, 2).

Posons :

$$n = \rho + 1 \qquad \rho : 0, 1, \ldots \qquad n : 1, 2, \ldots.$$

Ecrivons l'équation (II, 30) sous la forme :

$$\sqrt{m^2c^4-E^2} = \frac{nmc^2}{4\lambda}\left(\sqrt{1+\frac{8\lambda^2}{n^2}}-1\right)$$

A des termes de l'ordre de λ^3 près nous avons :

$$\sqrt{m^2c^4-E^2} = \frac{mc^2\lambda}{n}$$

et par conséquent à des termes de l'ordre de λ^4 près :

$$m^2c^4-E^2 = \frac{m^2c^4\lambda^2}{n^2}$$

d'où il vient finalement à des termes de l'ordre de λ^3 près

$$E = mc^2 - \lambda^2 \frac{mc^2}{2n^2}$$

ce qui établit le résultat énoncé puisqu'il coincide au terme additif mc^2 près avec (B, 12).

Le postulat de quantification que nous avons adopté conduit donc, pour $\rho \geqslant 0$, à des résultats tout-à-fait satisfaisants.

11. - L'état super-lié.

Pour $\rho = -1$ un fait nouveau se présente qui n'a pas d'analogue dans la théorie du potentiel central de Newton. En effet pour $\rho = -1$ l'équation (II, 24) donne

$$\mathcal{H}^2 = \frac{q}{2p}$$

Soit, compte tenu des définitions respectives :

$$E = \frac{1}{\sqrt{2}} mc^2$$

Cette valeur de E est la valeur de l'énergie qui entraînait un comportement à l'infini en r^{-2} pour le potentiel effectif (II, 19).

Nous devons faire deux remarques au sujet de cet état qu'on peut appeler super-lié.

a) La valeur de E correspondante ne dépend pas de λ. Ceci est probablement dû au fait que les valeurs de ρ précises, que l'on déduirait de la solution correcte du problème de raccordement des solutions aux voisinages des points singuliers, ne seraient pas entières mais differeraient des entiers pas des termes de l'ordre de λ^2. C'est ce qui arrive dans le problème de l'App. C pour lequel il existe aussi un état super-lié dont l'origine est similaire à l'origine de celui que nous discutons ici.

b) L'énergie de liaison qui est

$$L = mc^2 - E = \frac{\sqrt{2}-1}{\sqrt{2}} mc^2 \simeq 0,3\, mc^2$$

est d'un ordre de grandeur bien supérieur à celui qui correspond aux états newtonniens qui est de l'ordre de $\lambda^2 \frac{mc^2}{2}$.

On retrouve encore un décalage similaire dans l'ordre de grandeur des énergies de liaison des états liés correspondant au problème de l'App. C.

Conclusion.

Au point de vue physique ce chapitre comporte essentiellement deux résultats. Un résultat attendu : l'existence d'une série discrète d'états liés dont l'énergie de liaison est celle que l'on déduit de la théorie du potentiel central newtonnien. Et un résultat inattendu : l'existence d'un état super-lié. Nous tenons à dire que les doutes qu'on puisse avancer sur l'existence de cet état nous paraîssent tout-à-fait légitimes, beaucoup de travail formel restant à faire pour pouvoir établir son existence sur des bases plus solides. Cependant nous croyons qu'il y a deux arguments de plausibilité qui inclinent à croire à son existence.

Premièrement, l'analogie formelle entre le problème que nous avons envisagé et celui de l'App. C pour lequel, formellement au moins, l'existence d'un état super-lié est hors de doute.

Deuxièmement, parce qu'il serait vraiment surprenant que deux théories, celle de Schwarzschild et celle de Newton, dont les prévisions théoriques classiques peuvent diffèrer énormement pour des conditions initiales convenablement choisies, fussent tout à fait équivalentes du point de vue quantique.

L. Bel

CHAPITRE III : ONDES PLANES A L'INFINI

Introduction

Nous abordons dans ce chapitre le problème de la généralization, au ds^2 de Schwarzschild, de la notion d'onde plane familière, et combien utile, en Relativité Restreinte.

Ce problème est équivalent, dans un sens, au problème de la caractérization des faisceaux de particules d'épreuve à énergie et impulsion linéaire définies.

Nous proposons des définitions pour la notion d'onde plane à l'infini au sens géométrique d'abord et au sens physique ensuite. Nous établissons les étapes d'un programme qui devrait permettre, en principe, la construction d'ondes planes à l'infini au sens géométrique et réalisons effectivement ce programme à une approximation qui sera précisée. A cette approximation nous obtenons aussi des ondes planes à l'infini au sens physique.

Les résultats ici obtenus ont des rapports avec au moins deux problèmes physiques précis.

D'une part, le problème de la diffusion de deux particules à interaction purement gravitationnelle. Dans ce sens ce chapitre peut être considéré comme une prolongation du précédent.

D'autre part, avec le problème de la formulation du phénomène de l'Aberration et de l'effet Döppler dans l'espace-temps de Schwarzschild.

1. - Les ondes planes en Relativité Restreinte.

Une famille complète d'ondes planes en Relativité Restreinte est définie par la fonction

(III, 1) $$S(x^i, t \mid P_i, E) = -Et + P_i x^i$$

dépendant de quatre paramètres E et P_i que nous supposerons, pour nous restreindre aux cas d'intérêt physique plus immédiat, astreints à satisfaire les conditions

$$E^2 - c^2 \sum_i P_i^2 \geqslant 0 \qquad E > 0$$

L'object de ce chapitre est de généraliser la famille de fonctions S tout en conservant à la généralisation les propriétés qui font l'importance de la notion d'onde plane. Il importe donc de savoir qu'elles sont ces propriétés aussi bien au point de vue formel, ce qui nous donnera des indices sur la définition généralisée qu'il conviendra de retenir, qu'au point de vue de l'interprétation physique, ce qui précisera le domaine d'application des résultats ici obtenus.

Considérons le point de vue formel d'abord.

En Relativité Restreinte l'intégrale d'action d'une particule de masse m est

$$S = \int_{t_o}^{t} - mc^2 \sqrt{1 - \frac{v^2}{c^2}}\, dt$$

L'équation de Hamilton-Jacobi associée au système dynamique défini par l'intégrand ci-dessus est :

(III, 2) $$\Delta_1 S \equiv \left(\frac{\partial S}{\partial t}\right)^2 - c^2 \sum \left(\frac{\partial S}{\partial x^i}\right)^2 = m^2 c^4$$

et on constate que si

$$E^2 - c^2 \sum P_i^2 = m^2 c^4$$

la fonction S (III, 1) est une intégrale complète de (III, 2).

Une deuxième constatation est que :

(III, 3) $$\Delta_2 S \equiv \left(\frac{\partial^2 S}{\partial t^2}\right) - c^2 \sum \frac{\partial^2 S}{\partial x_i^2} = 0$$

De (III, 2) et (III, 3) il vient

$$(\Delta_2 + m^2 c^4) e^{-iS} = 0$$

ou encore, si l'on veut :

(III, 4) $$(\Delta_2 + \frac{m^2 c^4}{\hbar^2}) \psi = 0$$

avec

(III, 5) $$\psi = e^{-i\frac{S}{\hbar}}$$

Ces deux constatations formelles se rapportent aux deux aspects physiques de la notion d'onde plane que nous voulions dégager.

Premièrement, la fonction (III, 1) est une intégrale complète de (III, 2) certes, mais aussi une intégrale complète très particulière. En effet les constantes d'intégration E et P_i sont respectivement l'énergie et les composantes du moment linéaire. Elle décrit, donc, un faisceau de particules d'épreuve de masse m à énergie et impulsion linéaire définies.

Deuxièmement la fonction ψ (III, 5), étant solution de (III, 4) est la fonction d'onde, au sens de la mécanique ondulatoire, du faisceau de particules décrit par la fonction S .

2. - Définition d'onde plane à l'infini au sens géométrique.

Considérons une congruence de courbes $\Lambda(\vec{n})$ et la famille de droites $\Gamma(\vec{n})$ parallèles à un vecteur unitaire $\vec{n}$ donné. Nous dirons que $\Gamma(\vec{n})$ est la congruence asymptotique de $\Lambda(\vec{n})$ si :

a) on peut établir une correspondance biunivoque entre les éléments de $\Gamma(\vec{n})$ et $\Lambda(\vec{n})$ de sorte que les droites de $\Gamma(\vec{n})$ soient des asymptotes tangentes de leurs images par cette correspondance.

b) le point de tangence asymptotique pour chaque paire d'éléments homologues est situé du même côté par rapport au plan orthogonal à $\vec{n}$.

Définition. - Une onde monochromatique et plane à l'infini au sens géométrique, ou pour abréger, onde plane à l'infini, pour le ds^2 de Schwarzschild est toute fonction $S(\xi^j, t)$ satisfaisant aux trois propriétés suivantes:

1) est solution de l'équation

$$(\text{III}, 6) \qquad \Delta_1 S \equiv \frac{1}{\gamma}\left(\frac{\partial S}{\partial t}\right)^2 - c^2\gamma\left(\frac{\partial S}{\partial r}\right)^2 - \frac{c^2}{r^2}\left[\left(\frac{\partial S}{\partial \theta}\right)^2 + \frac{1}{\sin^2\theta}\left(\frac{\partial S}{\partial \varphi}\right)^2\right] = m^2c^4$$

2)

$$\frac{\partial S}{\partial t} = -E \qquad E^2 > m^2c^4 \,, \qquad E > 0$$

3) il existe un vecteur $\vec{n}$ tal que la congruence $\Lambda(n)$ de trajectoires orthogonales à la famille de surfaces $S(\xi^i, t) = $ cte pour t variable ait la congruence de droites $\Gamma(n)$ pour congruence asymptotique.

Soit S une fonction satisfaisant aux conditions de la définition précédente et soient α_i les composantes du vecteur $\vec{n}$ correspondant. La première condition exprime que la fonction S est une solution de l'équation de Hamilton-Jacobi (I, 20). La deuxième que le faisceau de particules d'eprèuve décrit par S est un faisceau de particules de même énergie E. Enfin la troisième condition exprime que chaque particule du faisceau a, au point de l'infini où la tangence avec l'élément correspondant de $\Gamma(\vec{n})$ a lieu, une impulsion linéaire

$$P_i = \eta\frac{\alpha_i}{c}\sqrt{E^2 - m^2c^4} \qquad \eta = \pm 1 \qquad i = 1, 2, 3$$

On peut donc dire d'une telle fonction S, si elle existe, qu'elle décrit un faisceau de particules de même masse à énergie-impulsion linéaire définie. Nous supposerons par la suite $\eta = +1$ avec quoi il s'agira d'un faisceau incident, la direction et le sens de l'incidence étant ceux du vecteur $\vec{n}$.

3. - Programme pour la construction effective.

a. - Nous avons déjà rencontré des solutions de l'équation (III, 6) au chap. I. En effet nous avons vu que quelles que soient les constantes E, M et P pour m qu'elles conduisent à des expressions réelles la fonction (I, 23) :

$$S = -Et + \frac{\varepsilon}{c}\int \frac{\Omega}{\sigma}\, dr + \gamma \int \sqrt{P^2 - \frac{M^2}{\sin^2\theta}}\, d\theta + M\varphi \tag{III, 7}$$

où :

$$\Omega = \sqrt{E^2 - \frac{c^2P^2\sigma}{r^2} - m^2c^4\sigma} \tag{III, 8}$$

est une solution de (III, 6). Nous allons nous en servir pour obtenir une nouvelle fonction S particulière satisfaisant aux conditions de la définition d'onde plane à l'infini avec quoi la preuve de l'existence sera faite.

Posons

$$M = 0$$

et supposons

$$E^2 - m^2c^4 \geqslant 0$$

La différentielle de la fonction (III, 7) compte tenu de M = 0 est

$$dS = -E\, dt + \frac{\varepsilon}{c\sigma}\, \Omega\, dr + P\, d\vartheta \tag{III, 9}$$

L'intégrale générale de cette équation décrit maintenant un faisceau de particules dont chacune des trajectoires est contenue dans un plan φ = cte et a pour équation dans ce plan (Voir Ch. I Sec. II)

$$\theta - \theta_0 = \varepsilon cP\int_{r_0}^{r} \frac{dr}{r^2\Omega}$$

b. - Considérons le faisceau de droites parallèles à l'axe z, $\Gamma(\vec{n})$:

$$r \sin\theta = s \qquad -\infty < s < \infty$$

$$\varphi = \varphi_o \qquad 0 \leq \varphi_o < 2\pi$$

et faisons correspondre à chaque valeur de s la valeur de P :

$$P = -\frac{s}{c}\sqrt{E^2 - m^2c^4} \tag{III, 10}$$

Faisons maintenant correspondre à chaque paire de valeurs (s, φ_o), c'est-à-dire à chaque droite du faisceau $\Gamma(\vec{n})$ la trajectoire de la particule d'épreuve d'équations au voisinage de $\theta = \pm\pi$ est :

$$\theta - \gamma\pi = -cP\int_{\infty}^{r} \frac{dr}{r^2\Omega} \equiv f(r, P, E) \qquad \varphi = \varphi_o \tag{III, 11}$$

P ayant au deuxième membre la valeur (III, 10) et γ étant le signe de s . Nous obtenons ainsi une congruence de courbes qui sont les trajectoires d'un faisceau de particules d'épreuve de masse m , énergie E et ayant au voisinage de $\pm\pi$ une vitesse radiale négative. Il est clair d'ailleurs que la correspondence envisagée peut être rendue biunivoque.

Chaque élément de $\Lambda(\vec{n})$ a un point à l'infini pour $\theta = \gamma\pi$. Les équations des asymptotes en ces points sont :

$$r \sin\theta = \lim_{r\to\infty} r^2 \frac{d\theta}{dr} \qquad \varphi = \varphi_o$$

Or de (III, 11) il vient :

$$\frac{d\theta}{dr} = -\frac{cP}{r^2\Omega}$$

et puisque, d'après (III, 8) et (I, 9)

$$\lim_{r\to\infty} \Omega = \sqrt{E^2 - m^2c^4}$$

nous pouvons écrire les équations des asymptotes sous la forme :

$$r \sin \theta = - \frac{cP}{\sqrt{E^2 - m^2 c^4}} \qquad \varphi = \varphi_o$$

ou encore d'après (III, 10)

$$r \sin \theta = s \qquad \varphi = \varphi_o$$

Ainsi chaque élément de $\Lambda(\vec{n})$ a pour asymptote l'élément homologue de $\Gamma(\vec{n})$ par la correspondance établie. En fait comme il est facile de le voir il y a même tangence asymptotique.

c. - Soit $P(r, \theta, E)$ la fonction implicite définie par la formule (III, 11):

$$\theta - \gamma \pi = f(r, P, E)$$

où r est supposé être plus grand que la plus grande racine de $\Omega^2 = 0$. La valeur de cette racine dépend évidemment de P et par conéquent de s.

Considérons maintenant de nouveau l'équation (III, 9)

$$dS = - Edt - \frac{1}{c\sigma} \Omega \, dr + Pd\theta \tag{III, 12}$$

mais supposons ici que P est la fonction $P(r, \theta, E)$ que nous venons de définir. Cette équation est encore complètèment intégrable. En effet, d'après la définition de la fonction P ses dérivées par rapport à r et à θ seront telles que :

$$\frac{\partial f}{\partial P} \frac{\partial P}{\partial \theta} = 1 \quad , \quad \frac{\partial f}{\partial r} + \frac{\partial f}{\partial P} \frac{\partial P}{\partial r} = 0$$

d'où il vient en éliminant $\frac{\partial f}{\partial P}$:

$$\frac{\partial P}{\partial r} = - \frac{\partial f}{\partial r} \frac{\partial P}{\partial \theta}$$

ou encore d'après la définition de f :

$$\frac{\partial P}{\partial r} = \frac{cP}{r^2 \Omega} \frac{\partial P}{\partial \vartheta}$$

ce qui compte tenu de :

$$\frac{\delta}{\delta \theta}\left(\frac{\Omega}{c\sigma}\right) \equiv \frac{1}{c\sigma} \frac{\partial \Omega}{\partial P} \frac{\partial P}{\partial \vartheta} = \frac{cP}{r^2 \Omega} \frac{\partial P}{\partial \theta}$$

est la condition nécessaire et suffisante pour que l'équation (III, 12) soit complètément intégrable.

Soit :

$$S = -Et + W(r, \vartheta \,/\, E) \tag{III, 13}$$

l'intégrale générale de (III, 12). Elle satisfait toutes les conditions de la définition d'onde plane à l'infini. En effet d'après (III, 12) les conditions 1) et 2) sont satisfaites; la condition 3) aussi puisque la congruence de trajectoires orthogonales n'est autre que la congruence Λ $(\vec{n})$ et que celle-ci satisfait aux conditions requises par construction.

Physiquement la fonction (III, 13) décrit un faisceau incident de particules d'épreuve de masse m, énergie E et impulsion à l'incidence

$$P_1 = 0 \qquad P_2 = 0 \qquad P_3 = \frac{\sqrt{E^2 - m^2c^4}}{c}$$

4. - Famille à quatre paramètres d'ondes planes à l'infini.

Nous nous proposons de montrer comment à partir de la fonction (III, 13) on pourrait obtenir une famille à quatre paramètres d'ondes planes.

Supposons que nous effectuons le changement de coordonnées

$$r = \sqrt{x^2+y^2+z^2}, \quad \theta = \text{arc tn}\, \frac{\sqrt{x^2+y^2}}{z}, \quad \varphi = \text{arc tn}\, \frac{y}{x}, \quad t = t \tag{III, 14}$$

et puis le changement de coordonnées

$$x^i = R^i_{j'} x^{j'} \qquad t' = t \qquad x' = x \qquad x^2 = y \qquad x^3 = z \tag{III, 15}$$

$R^i_{j'}$ étant une matrice orthogonale quelconque.

Posons

$$\alpha_{j'} = R^3_{j'}$$

Nous aurons donc :

(III, 16) $$\sum \alpha^2_{j'} = 1$$

D'autre part :

$$r = \sqrt{\sum (x^i)^2} = \sqrt{\sum (x^{i'})^2}$$

$$\theta = \text{arc tn} \frac{\sqrt{x^2+y^2}}{z} = \text{arc tn} \frac{\sqrt{r^2-z^2}}{z} = \text{arc tn} \frac{\sqrt{(x^{i'})^2 - (\alpha_{j'}x^{j'})^2}}{\alpha_{j'}x^{j'}}$$

et par conséquent après les changements de coordonnées envisagés, la fonction (III, 13) sera de la forme

(III, 17) $$S = -E't' + W(x^{i'}/ P_{i'})$$

où :

(III, 18) $$P_{i'} = \alpha^{i'} \frac{\sqrt{E'^2-m^2c^4}}{c} \qquad E' = E$$

Il est clair d'après la signification géométrique des changements de coordonnées effectués que la fonction (III, 17) est une onde plane à l'infini dont la direction d'incidence est celle du vecteur $\vec{n}$ qui a pour composantes

$$n_{i'} = \alpha_{i'}$$

Nous pouvons considérer le système de coordonnées fixe et les $\alpha_{i'}$ arbitraires, mais satisfaisant la condition (III, 16). Ou même, nous pouvons considérer E' et $P_{i'}$ arbitraires, mais satisfaisant la condition

$$E'^2 - c^2 \sum P^2_{i'} \geqslant 0$$

et dans ce cas chacune des fonctions (III, 17) sera solution de l'équation

$$\Delta_1 \equiv m^2 c^4$$

où maintenant m n'est pas un paramètre donné d'avance mais qui sera tel que

$$m^2 c^4 = E'^2 - c^2 \sum P^2_{i'}$$

C'est dans ce sens que nous dirons que (III, 17) définit une famille à quatre paramètres d'ondes planes à l'infini.

5. - Construction effective approchée.

a. - Les paragraphes précédents ne contiennent en fait que la preuve de l'existence d'ondes planes à l'infini et les étapes d'un programme qui devrait en principe en permettre la construction effective. La réalisation de ce programme s'avère cependant difficile, ou même peut-être impossible, pour des raisons de calcul.

Nous nous proposons ici de suivre pas à pas les étapes du programme décrit, nous bornant à un certain ordre d'approximation. D'une manière plus précise : nous développerons en série systématiquement toute les fonctions qui dépendent de $\mu = \frac{kM}{c^2}$ par rapport à ce paramètre et négligerons les termes d'ordre supérieur à μ. Nous pouvrions donc considérer les résultats obtenus comme corrects pour M suffisamment petit. Mais c'est une autre intèrprétation qui nous intèresse surtout. Nous pouvons toujours écrire $\mu = \frac{\mu}{r} r$ et considérer $\frac{\mu}{r}$ comme l'infiniment petit principal ce qui revient à multiplier par r le coefficient du développement. Dans ce cas les résultats obtenus pourront être considérés corrects pour r suffisamment grand.

b. - Posons

$$u = \frac{1}{r}, \quad \chi = cP, \quad \zeta^2 = E^2 - m^2 c^4 > 0, \quad \zeta > 0$$

L'équation (III, 11) peut s'écrire :

$$\theta - \gamma\pi = \chi \int_0^u \frac{du}{\Omega}$$

Le développement de Ω^{-1} à des termes de l'ordre de μ^2 près, compte tenu de (III, 8), et :

(III, 19) $$\sigma = 1 - 2\mu u$$

est

$$\frac{1}{\Omega} = \frac{1}{\sqrt{\zeta^2 - \chi^2 u^2}} + \mu \left[\frac{u}{\zeta^2 - \chi^2 u^2} - \frac{uE^2}{(\zeta^2 - \chi^2 u^2)^{3/2}} \right]$$

ce qui donne après intégration :

$$\theta - \gamma\pi = \arcsin \frac{u\chi}{\zeta} + \frac{\mu}{\chi} \left[\frac{E^2}{\zeta} - \frac{E^2}{\sqrt{\zeta^2 - u^2\chi^2}} + \zeta - \sqrt{\zeta^2 - u^2\chi^2} \right]$$

et d'où, en prenant le sinus des deux membres, on tire au même ordre d'approximation :

$$-\sin\theta = \frac{u\chi}{\zeta} + \frac{\mu}{\chi} \sqrt{1 - \frac{u^2\chi^2}{\zeta^2}} \left[\frac{E^2}{\zeta} - \frac{E^2}{\sqrt{\zeta^2 - u^2\chi^2}} + \zeta - \sqrt{\zeta^2 - u^2\chi^2} \right]$$

et de même en prenant le cosinus :

$$-\cos\theta = \sqrt{1 - \frac{u^2\chi^2}{\zeta^2}} - \frac{u\mu}{\zeta} \quad \frac{E^2}{\zeta} - \frac{E^2}{\sqrt{\zeta^2 - u^2\chi^2}} + \zeta - \sqrt{\zeta^2 - u^2\chi^2}$$

Par conséquent nous pouvons écrire :

$$\sin\theta = -\frac{u\chi}{\zeta} + \frac{\mu}{\chi} \left[\frac{E^2\cos\theta}{\zeta} + \frac{E^2}{\zeta} + \zeta\cos\theta + \frac{\zeta^2 - u^2\chi^2}{\zeta} \right]$$

D'où il vient en multipliant les deux membres par $\frac{u\chi}{\zeta}$ et en groupant les termes :

$$\frac{u\chi}{\zeta} \sin\theta = -\frac{u^2\chi^2}{\zeta^2} + u\mu \left[\left(\frac{E^2}{\zeta^2} + 1\right)(\cos\theta + 1) - \frac{u^2\chi^2}{\zeta^2} \right]$$

Posons :

$$\frac{u\chi}{\zeta} = -\sin\theta + \mu R$$

et négligeons les termes de l'ordre de μ^2 on supérieurs. Nous obtenons:

$$-\sin^2\theta + \mu R\sin\theta = -\sin^2\theta + 2\mu R\sin\theta + u\mu\left[(\frac{E^2}{\zeta^2}+1)(\cos\theta+1)-\sin^2\theta\right]$$

et par conséquent :

$$R = -\frac{u}{\sin\theta}\left[(\frac{E^2}{\zeta^2}+1)(\cos\theta+1)-\sin^2\theta\right]$$

et finalement

$$\chi = -\frac{\zeta}{u}\sin\theta - \frac{\zeta\mu}{\sin\theta}\left[(\frac{E^2}{\zeta^2}+1)(\cos\theta+1)-\sin^2\theta\right]$$

ou :

$$P = -\frac{\zeta r}{c}\sin\theta - \frac{\mu}{c\zeta\sin\theta}\left[(E^2+\zeta^2)(\cos\theta+1)-\sin^2\theta\right] \quad \text{(III, 20)}$$

c. - Les deux premiers termes du développement de $\frac{\Omega}{\sigma}$ sont, compte tenu de (III, 8), (III, 19) et l'expression de P ci-dessus :

$$\frac{\Omega}{\sigma} = -\zeta\cos\theta + \frac{\mu}{r\zeta}(E^2+\zeta^2) \quad \text{(III, 21)}$$

La substitution de (III, 20) et (III, 21) dans (III, 12) donne :

$$dS = -Edt + \left[\frac{\zeta}{c}\cos\theta - \frac{\mu}{cr\zeta}(E^2+\zeta^2)\right]dr +$$

$$-\left\{\frac{\zeta r}{c}\sin\theta + \frac{\mu}{c\zeta\sin\theta}\left[(E^2+\zeta^2)(\cos\theta+1) - \sin^2\theta\right]\right\}d\theta$$

et l'intégrale générale de cette équation est :

$$S = -Et + \frac{\zeta}{c}r\cos\theta - \frac{\mu}{c\zeta}(E^2+\zeta^2)\mathrm{Log}(r\sin^2\frac{\theta}{2}) + \cos\theta = \text{cte.} \quad \text{(III, 22)}$$

Elle correspond à l'approximation envisagé à la fonction (III, 13). Négligeant les termes d'ordre supérieur a μ elle est solution de

(III, 23) $$\Delta_1 S = m_2 c^4$$

Ce n'est pas tout-à-fait évident du fait que l'opérateur Δ_1 contient lui-même des termes en μ mais peut être vérifié directement.

d. - Faisons les changements de coordonnées (III, 14) et (III, 15). Nous avons :

$$r \cos\theta = z = \alpha_{j'} x^{j'} \qquad \cos\theta = \frac{z}{r} = \frac{\alpha_{j'} x^{j'}}{r'} \qquad r' = \sqrt{\sum (x^{i'})^2}$$

$$r \sin^2 \frac{\theta}{2} = \frac{r(1-\cos\theta)}{2} = \frac{r-z}{2} = \frac{r' - \alpha_{j'} x^{j'}}{2}$$

et par conséquent l'expression de la fonction (III; 17) avec les notations (III, 18) est

$$S = -E't' + P_{j'} x^{j'} - \frac{\mu}{c\zeta}\left[(E^2+\zeta^2)\operatorname{Log}\frac{1}{2}\left(r' - \frac{c}{\zeta}P_{j'}x^{j'}\right) + \frac{c}{\zeta r'}P_{j'}x^{j'}\right]$$

où on reconnaît dans la partie principale l'expression en coordonnées rectilignes des fonctions des ondes planes incidentes dont la direction de propagation est celle du vecteur $\vec{n}$ de composantes $n_{j'} = \alpha_{j'}$.

6. - Définition d'onde plane à l'infini au sens physique.

Nous proposons la définition suivante :

Définition. - Une onde monochromatique et plane à l'infini au sens physique est toute fonction ψ, solution de l'équation

1) $$\left(\Delta_2 + \frac{m^2 c^4}{2}\right)\Psi = 0$$

de la forme :

2) $$\psi = A\, e^{-i\frac{S}{\hbar}}$$

S étant une onde monochromatique plane à l'infini au sens géométrique. Δ_2 est le laplacien intrinsèque sur les fonctions.

Dans ce qui suit nous prouvons que si :

$$A = 1$$

et S est la fonction (III, 17), la fonction Ψ correspondante satisfait, à l'ordre d'approximation envisagé au paragraphe précédent, aux conditions de la définition ci-dessus. Elle satisfait évidemment à la condition 2) et pour prouver qu'elle satisfait à la condition 1) il suffit de le prouver pour la fonction S (III, 13).

De

$$\Delta_2 \Psi = \Delta_2 (e^{-\frac{iS}{\hbar}}) = -(\frac{1}{\hbar^2} \Delta_1 S + \frac{i}{\hbar} \Delta_2 S) e^{-i\frac{S}{\hbar}}$$

et de (III, 23), il résulte que :

$$(\Delta_2 + \frac{m^2c^4}{\hbar^2}) \Psi = -\frac{i}{\hbar} \Delta_2 S \, e^{-i\frac{S}{\hbar}} \quad \text{(III, 24)}$$

Calculons $\Delta_2 S$. De

$$\Delta_2 S = \frac{1}{\sqrt{|g|}} \frac{\partial}{\partial \xi^\alpha} (\sqrt{|g|}\, g^{\alpha\beta} \frac{\partial S}{\partial \xi^\beta}),$$

de (I, 11) et de (III, 22) il vient :

$$\Delta_2 S = \frac{-c}{r^2 \sin\theta} \left\{ \frac{\partial}{\partial r} \left[r^2 \sin\theta \, (1 - \frac{2\mu}{r})(\zeta \cos\theta - \frac{\mu}{r\zeta}(E^2 + \zeta^2)) \right] - \frac{\partial}{\partial \theta} \left[\sin\theta \, (\zeta r \sin\theta + \frac{\mu}{\zeta \sin\theta} ((E^2 + \zeta^2)(\cos\theta + 1) - \sin^2\theta)) \right] \right\}$$

et il suffit de développer et négliger les termes en μ^2 pour obtenir à l'approximation envisagée

$$\Delta_2 S = 0$$

ce qui entraîne d'après (III, 24)

$$(\Delta_2 + \frac{m^2c^4}{\hbar^2}) \Psi = 0 \quad \text{avec} \quad \Psi = e^{-i\frac{S}{\hbar}} \quad \text{(III, 25)}$$

comme nous avions annoncé.

Remarque I. - Les résultats que nous avons exposés ici ont été utilisés pour formuler correctement les phénomènes de l'Aberration et l'effet Döppler par A. Montserrat à qui est dû d'ailleurs l'essentiel des calculs approchés. L'utilité de ces résultats pour étudier les phénomènes mentionnés tient au fait qu'ils permettent de caractériser avec précision le rayonnement émis par un object lumineux situé à l'infini. Plus précisément, ce rayonnement est décrit dans le travail de A. Montserrat par une onde monochromatique et plane à l'infini correspondant à $m = 0$.

Remarque II. - En ce qui concerne le problème de la diffusion des particules d'épreuve dans un champ de Schwarzschild et les résultats de ce chapitre, le rapport s'établit évidemment à travers l'équation (III, 25) qui est l'équation d'onde que nous avons déjà utilisée au chapitre II pour étudier les états liés. La forme de la fonction Ψ (III, 25) et le fait de satisfaire à l'approximation envisagée à l'équation (III, 25) en font la partie principale à l'infini incident d'un problème de diffusion correspondant à une énergie et direction d'incidence bien définies. Analoguement à ce qui arrive pour un potentiel newtonnien l'onde incidente est distorsionnée. Cette distorsion dans notre cas est le terme en μ de la formule (III, 22).

CHAPITRE IV :

DEUX ASPECTS PARTIELS DE LA GENERALISATION DU GROUPE DE LORENTZ INHOMOGENE.

Introduction.

Ce chapitre s'inspire de la conviction qu'en Relativité Générale, dans tout domaine extérieur, tout comme en Relativité Restreinte, il convient d'introduire deux notions fondamentales. Premièrement la notion d'observateur privilégié. Deuxièment la notion d'ensemble de transformation définissant la correspondance entre les coordonnées d'un même évènement par rapport à deux observateurs privilégiés.

La trajectoire de tout observateur pouvant être identifiée à une trajectoire spatio-temporelle orientée dans le temps, le premier problème revient donc à distinguer dans tout domaine extérieur des trajectoires temporelles particulières.

Le premier point de vue adopté ici est que toute géodésique orientée dans le temps peut être identifiée à la trajectoire d'un observateur privilégié qui généralise la notion d'observateur galiléen de la Relativité Restreinte. Ce n'est pas un point de vue nouveau. Il a été souvent exprimé, surtout dans le passé, mais sans beaucoup de conviction semble-il puisque à notre connaissance un seul article [5] à été consacré aux éxigences qu'un tel point de vue comporte.

Le deuxième point de vue est que dans les cas particuliers où la métrique admet un groupe d'isométries à un paramètre à trajectoires orientées dans le temps, chaque trajectoire du groupe peut être identifiée à la trajectoire d'un observateur privilégié qui est immobile par rapport aux sources. Ceci est, croyons-nous, couramment admis.

Le cas qui nous intéresse ici est le cas du ds^2 de Schwarzschild, qui est un ds^2 statique admettant par conséquent un groupe d'isométries à un paramètre à trajectoires orientées dans le temps. Nous distinguerons donc deux grandes classes d'observateurs privilégiés. Ceux que nous désignerons par 0, dont la trajectoire est une trajectoire du groupe indiqué, et ceux que nous désignerons par 0', dont la trajectoire est une géodésique orientée dans le temps.

Nous parlerons aussi d'un observateur C que nous supposerons être à l'origine des coordonnées polaires. Ce n'est que la convention habituelle pour nous référer à la classe 0 dans son ensemble, étant entendu que si $\vec{a}$ est le vecteur position d'un observateur de la classe 0 et $(\vec{r}, t)$ sont les coordonnées d'un évènement P par rapport à l'origine, les coordonnées de P par rapport à 0 sont par définition $(\vec{r} - \vec{a},\ t)$.

Les problèmes que nous abordons ici ne sont en fait que deux aspects partiels du problème général de définir la correspondance entre les coordonnées d'un même évènement par rapport à deux observateurs privilégiés.

Dans la première section nous nous bornons à considérer des observateurs de la classe 0' dont les trajectoires d'espace sont radiales et des évènements qui ont lieu sur le rayon de la trajectoire correspondante. Les résultats sont donnés sous forme d'algoritme permettant de calculer les fonctions de transformation des coordonnées d'un évènement P référé à un observateur de la sous-classe de 0' envisagée, et référé à C .

Dans la deuxième section nous considérons l'ensemble de la classe 0' ainsi que C mais par contre nous nous limitons au premier ordre d'approximation dans un sens qui sera précisé.

Il est clair qu'ayant les formules de transformation entre C et un observateur quelconque 0' nous pourrions en principe obtenir les formules de transformation entre un couple quelconque d'observateurs 0' par élimination, entre les formules déjà obtenues, des coordonnées relatives à C . C'est dans ce sens que strictement parlant les résultats ici obtenus sont un premier pas vers la généralisation du groupe de Lorentz inhomogène. Mais, tout en continuant à nous référer uniquement aux transformations entre C et 0', il est un autre sens dans lequel il est légitime de parler de généralisation du groupe de Lorentz inhomogène. En effet, ces fonctions de transformation entre C et 0' que nous obtenons se réduisent aux transformations définissant le groupe de Lorentz inhomogène si dans elle nous supposons que la masse centrale M est nulle. C'était évidemment une condition nécessaire.

Nous définissons, dans la première section, ce qu'on pourrait appeler "transformations spéciales de Newton" qui avec les transformations que nous obtenons plus tard et qu'on pourrait appeler "transformations spéciales de Schwarzschild" permettent de présenter nos résultats sous la forme du diagramme complet :

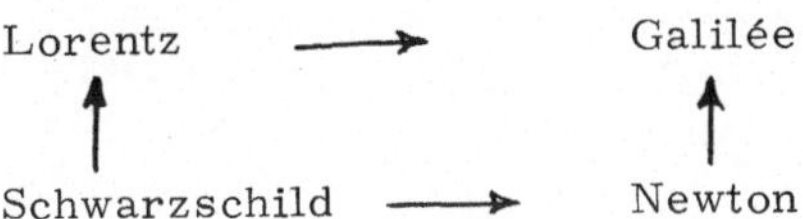

où une flèche horizontale veut dire que dans les formules de transformation correspondantes on a fait tendre c vers infini, et où une flèche verticale veut dire qu'on a fait tendre la constante de gravitation k , on ce que revient au même M , vers zéro.

Les résultats de la deuxième section réalisent aussi ce diagramme mais à l'approximation envisagé ce n'est qu'un résultat trivial et nous n'en reparlerons plus.

L. Bel

SECTION I :

GENERALISATION DU GROUPE DE LORENTZ INHOMOGENE SPECIAL A LA REDUCTION UNIDIMENSIONELLE RADIALE DU ds^2 DE SCHWARZSCHILD

1. - "Les transformations spéciales de Newton" .

Considérons une masse M à l'origine r=0 et une particule d'épreuve de masse m , soumise uniquement au champ de gravitation newtonnien de M, que nous identifierons à un observateur 0' en chute libre. Nous supposons la trajectoire de 0' radiale et par conséquent, abstraction faite, ce que nous fairons dans la suite de cette section, des coordonnées θ et φ, l'équation différentielle du mouvement de 0' est :

$$\frac{d^2r}{dt^2} = -\frac{kM}{r^2} \quad .$$

Plus précisément la trajectoire de 0', et par conséquent l'observateur 0' lui-même, sera caractérisée par la solution de cette équation correspondant aux conditions initiales r_o, t_o et W, r_o étant la position initiale de 0' à l'instant t_o et W étant l'intégrale première

$$W = \frac{1}{2}\dot{r}^2 - \frac{kM}{r} = \frac{1}{2}\dot{r}_o^2 - \frac{kM}{r_o}$$

c'est-à-dire l'énergie au facteur m près. Nous écrirons symboliquement $0'(P_o, W)$ pour désigner l'observateur correspondant.

Soient r et t les coordonnées d'un évènement P , suffisamment proche de P_o , par rapport à C , observateur fixe à r=0 . Nous nous proposons de déterminer les fonctions de transformation donnant les coordonnées r' et t' de ce même évènement par rapport à $0'(P_o, W)$.

Il est conforme au caractère absolu du temps dans la théorie de Newton de poser

(IV, 1) $$t'-t'_o = t-t_o$$

t'_o étant le temps pour 0' correspondant au temps t_o de C . Et il est conforme au caractère absolu de l'espace de poser, si on suppose que les directions r croissante et r' croissante coincident :

(IV, 2) $$r' = r-r_1$$

où :

$$r_1 = r(t-t_o, r_o, W)$$

$r(t-t_o, r_o, W)$ étant la solution de l'équation différentielle du mouvement correspondant aux conditions initiales qui caractérisent $0'(P_o, W)$. Cette fonction est donnée, comme on sait, sous forme implicite par l'équation :

$$t-t_o = \varepsilon \int_{r_o}^{r_1} \frac{dr}{\sqrt{2(W+\frac{kM}{r})}} \qquad \varepsilon = \frac{\dot{r}}{|r|} .$$

En développant le deuxième membre en série jusqu'au terme en $(r_1-r_o)^2$ il vient :

$$t-t_o = \frac{\varepsilon}{\sqrt{2(W+\frac{kM}{r_o})}} (r_1-r_o) + \frac{\varepsilon\, kM}{2r_o^2\left[2(W+\frac{kM}{r_o})\right]^{3/2}} (r_1-r_o)^2 .$$

En inversant cette série nous obtenons jusqu'au terme en $(t-t_o)^2$:

$$r_1-r_o = \varepsilon\sqrt{2(W+\frac{kM}{r_o})}(t-t_o) - \frac{kM}{2r_o^2}(t-t_o)^2 .$$

Or (IV, 2) peut s'écrire

$$r' = r - r_o - (r_1-r_o)$$

et, par conséquent, nous obtenons pour r' à l'approximation envisagée

(IV, 3) $$r' = r - r_o - \varepsilon \sqrt{2\left(W + \frac{kM}{r_o}\right)}(t-t_o) + \frac{kM}{2r_o^2}(t-t_o)^2 .$$

Cette formule de transformation est une généralisation de la formule de transformation spatiale du groupe de Galilée à laquelle elle se réduit si $kM = 0$. Ceci est vrai, à l'approximation envisagée, par inspection de la formule (IV, 3), mais c'est vrai aussi à tous les ordres par construction, comme on peut le constater facilement.

2. - Le groupe de Lorentz spécial inhomogène.

a. - Considérons les deux fonctions de transformation, relatives à deux observateurs galiléens 0 et 0', définissant le groupe de Lorentz spécial inhomogène :

(IV, 4a) $$t'-t'_o = \frac{t-t_o - v/_{c^2}(r-r_o)}{\sqrt{1 - \frac{v^2}{c^2}}}$$

(IV, 4b) $$r' = \frac{r-r_o - v(t-t_o)}{\sqrt{1 - \frac{v^2}{c^2}}} .$$

L'interprétation de ces formules est fournie par l'interprétation des différents arguments qui y figurent. Ainsi si r et t sont les coordonnées d'un évènement P par rapport à 0 , et r_o est la position initiale à l'instant t_o de 0 , qui a une vitesse v par rapport à 0 , r' et t' sont les coordonnées du même évènement par rapport à 0'. Cette interprétation suppose en outre que t'_o est le temps pour 0' qui correspond au temps t_o de 0 et que les directions positives des axes r et r' coincident.

Nous nous proposons pour l'instant d'établir un certain nombre de résultats concernant les formules (IV, 4) aussi bien au point de vue formel qu'au point de vue de certaines interprétations dont elles sont susceptibles. Ces résultats nous suggéreront dans quel sens il convient de tenter la généralisation au cas de la réduction unidimensionnelle radiale du ds^2 de Schwarzschild.

b. - Posons :

$$E = \frac{-1}{\sqrt{1-\frac{v^2}{c^2}}} \qquad N = \frac{v}{\sqrt{1-\frac{v^2}{c^2}}}$$

d'où il vient :

$$v = \frac{-\varepsilon c}{E}\sqrt{E^2-1} \qquad v = \frac{Nc}{\sqrt{N^2+c^2}} \qquad \varepsilon = \frac{v}{|v|} \tag{IV, 5}$$

et :

$$N = \varepsilon c \sqrt{E^2-1} \qquad E = -\frac{1}{c}\sqrt{N^2+c^2} \; .$$

Avec ces notations nous pouvons écrire les formules (IV, 4) sous la forme :

$$\tau(P|E|P_o) \equiv t'-t'_o = -E(t-t_o) - \frac{\varepsilon}{c}\sqrt{E^2-1}\,(r-r_o) \tag{IV, 6a}$$

$$\mathcal{H}(P|E|P_o) \equiv r' = -\varepsilon c \sqrt{E^2-1} - E\,(r-r_o) \tag{IV, 6b}$$

ou encore pour cette dernière :

$$\mathcal{H}(P|N|P_o) \equiv r' = -N(t-t_o) + \frac{1}{c}\sqrt{N^2+c^2}\,(r-r_o) \; . \tag{IV, 7}$$

Considérons les deux intégrales définissant respectivement le temps propre et la distance propre

$$\tau_{P_o}^{P_1} = \int_{t_o}^{t_1} \sqrt{1-\frac{v^2}{c^2}}\,dt \qquad \mathcal{H}_{P_1}^{P} = \int_{t_1}^{t} \varepsilon\sqrt{u^2-c^2}\,dt \qquad u = \frac{dr}{dt} \; . \tag{IV, 8a, b}$$

Le ε de la deuxième intégrale correspondra dans les considérations qui suivent à la convention adoptée suivant laquelle les directions positives de r et r' coincident. Ces deux intégrales définissent des problèmes variationnels pour lesquels les équations de Hamilton-Jacobi respectives sont :

$$\text{(IV, 9a, b)} \qquad \left(\frac{\partial \tau}{\partial t}\right)^2 - c^2\left(\frac{\partial \tau}{\partial r}\right)^2 = 1 \qquad \frac{1}{c^2}\left(\frac{\partial \mathcal{H}}{\partial t}\right)^2 - \left(\frac{\partial \mathcal{H}}{\partial r}\right)^2 = -1 \quad .$$

On vérifie immédiatement que (IV, 6a) et (IV, 7) sont respectivement des intégrales complètes des équations précédentes. En outre de :

$$\frac{\partial \tau}{\partial t}(P|E|P_o) = -E \qquad \frac{\partial \mathcal{H}}{\partial t}(P|N|P_o) = -N$$

il résulte que E et N sont précisément les Hamiltoniens respectifs associés aux intégrales (IV, 8). E est, au facteur $-mc^2$ près, l'énergie d'une particule de masse m .

Les transformations du groupe de Lorentz spécial sont donc des intégrales complètes des équations de Hamilton-Jacobi associées aux intégrales définissant le temps propre et la distance propre, les constantes d'intégration étant les Hamiltoniens respectifs.

c. - Ecrivons symboliquement l'équation de l'extrémale de (IV, 8a) correspondant aux conditions initiales P_o, E sous la forme :

$$G(A|E|P_o) = 0$$

A étant le point variable sur l'extrémale. D'après le résultat ci-dessus et les résultats classiques de mécanique analytique il y a donc équivalence entre les deux relations :

$$\text{(IV, 10)} \qquad G(A|E|P_o) = 0 \Longleftrightarrow \frac{\partial \tau}{\partial E}(A|E|P_o) = 0 \quad .$$

Calculons d'autre part la dérivée totale par rapport à t de $\mathcal{H}(A|E|P_o)$ le long de $G(A|E|P_o) = 0$. De (IV, 6b) il vient :

$$\frac{d\mathcal{H}}{dt}(A|E|P_o) = -\,\varepsilon\, c\sqrt{E^2-1} - Ev$$

soit d'après (IV, 5) :

$$\frac{d\mathcal{H}}{dt}(A|E|P_o) = 0 \quad .$$

Comme on a aussi :

$$\mathcal{H}(P_o|E|P_o) = 0$$

il résulte qu'il y a donc équivalence entre les deux relations

$$G(A|E|P_o) = 0 \Longleftrightarrow \mathcal{H}(A|E|P_o) = 0 \quad . \tag{IV, 11}$$

Evidemment (IV, 10) et (IV, 11) donnent dans les deux cas :

$$G(A|E|P_o) = 0 \Longleftrightarrow r - r_o - v(t-t_o) = 0 \quad .$$

Les deux fonctions de transformation (IV, 6) définissent donc à travers (IV, 10) et (IV, 11) deux expressions équivalentes de l'intégrale générale des équations différentielles des extrémales de (IV, 8a).

Soit $G(A|E|P_o) = 0$ l'équation de la trajectoire d'un observateur galiléen $0'(P_o, E)$. Pour bien comprendre l'intêret des résultats ci-dessus il convient d'insister sur le fait que, alors que dans (IV, 6) r et t sont les coordonnées d'un évènement quelconque, dans (IV, 10) et (IV, 11) r est la position de $0'$ par rapport à 0 à l'instant t. Cette ambivalence d'interprétations des formules (IV, 6) vient de l'ambivalence d'interprétations de l'intégrale (IV, 8a) qui est aussi bien l'intégrale de temps propre d'un observateur galiléen que l'intégrale d'action, au facteur $-mc^2$ près, des particules libres.

d. - Soit P un évènement quelconque et considérons la variété $\mathcal{T}(A|E|P_o) = \mathcal{T}(P|E|P_o)$. Cette variété étant une intégrale complète de (IV, 9a) correspondant à la constante E elle est une variété tranversale à l'extrémale $G(A|E|P_o) = 0$. Soit P_1 le point d'intersection des deux. P_1 étant sur la variété transversale nous aurons $\mathcal{T}(P_1|E|P_o) = \mathcal{T}(P|E|P_o)$ et

P_1 et P_o étant sur la même extrémale, d'après (IV, 6a) et le théorème du ch. I. Sect. I, il vient :

$$\text{(IV, 12a)} \qquad t'-t'_o = \tau_{P_o}^{P_1} = \int_{t_o}^{t_1} \sqrt{1 - \frac{v^2}{c^2}}\, dt$$

l'intégrale étant calculée le long de $G(A|E|P_o) = 0$.

En inversant les rôles des intégrales (IV, 8a) et (IV, 8b) un raisonnement identique permet d'obtenir le résultat correspondant :

$$\text{(IV, 12b)} \qquad r' = \mathcal{H}_{P_1}^{P} = \int_{t_1}^{t} \varepsilon \sqrt{u^2 - c^2}\, dt$$

l'intégrale étant calculée le long de $\tau(A|E|P_o) = \tau(P|E|P_o)$.

Note. - D'après la remarque II de la fin de Chap. I, sect. II, on peut résumer les résultats précédents en termes des métriques :

$$d\tau^2 = dt^2 - \frac{1}{c^2} dr^2 \qquad d\mathcal{H}^2 = dr^2 - c^2 dt^2$$

de la manière suivante : les variétés $\tau(A|E|P_o)$ = cte sont d'une part des variétés orthogonales à la famille de géodésiques $G(A|E|P_o) = 0$, d'autre part des géodésiques de $d\mathcal{H}^2$. (IV, 12) montrent donc que $t'-t'_o$ et r' sont les coordonnées géodésiques de P par rapport à 0'.

3. - Généralisation du groupe de Lorentz spécial inhomogène

Considérons la réduction unidimensionnelle radiale du ds^2 de Schwarzschild :

$$\text{(IV, 13)} \qquad ds^2 = \sigma\, dt^2 - \frac{1}{\sigma c^2} dr^2 \qquad \sigma = 1 - \frac{2\mu}{r} \qquad \mu = \frac{kM}{c^2} .$$

Les intégrales de temps propre et distance propre correspondantes sont

$$\text{(IV, 14a, b)} \quad \tau_{P_o}^{P_1} = \int_{t_o}^{t_1} \sqrt{\sigma - \frac{v^2}{\sigma c^2}}\, dt , \quad \mathcal{H}_{P_1}^{P} = \int_{t_1}^{t} \varepsilon \sqrt{\frac{u^2}{\sigma} - c^2 \sigma}\, dt, \; u = \frac{dr}{dt}, \; t = \frac{v}{|v|}$$

Soit $G(A|E|P_o) = 0$ l'équation de l'extrémale de (IV, 14a) correspondant aux conditions initiales (P_o, E), E étant l'Hamiltonien

(IV, 15) $$E = \frac{-\sigma}{\sqrt{\sigma - \frac{v^2}{\sigma c^2}}}$$

qui est une intégrale première des équations differentielles des extrémales de (IV, 14a). Et soit $\tau(A|E|P_o)$ = cte la famille de transversales correspondant à la même constante E . Considérons un point quelconque P et la transversale de la famille précédente qui passe par P . Son équation est $\tau(A|E|P_o) = \tau(P|E|P_o)$. Elle coupera l'extrémale $G(A|E|P_o) = 0$ en un point P_1. Nous nous proposons de résoudre le problème suivant :

Déterminer les fonctions :

(IV, 16a, b) $$t'-t'_o \equiv \tau_{P_o}^{P_1} \qquad r' \equiv \mathcal{H}_{P_1}^{P}$$

la première intégrale étant calculée le long de $G(A|E|P_o) = 0$, la seconde le long de $\tau(A|E|P_o) = \tau(P|E|P_o)$.

Si $0'(P_o$, E) est l'observateur de 0' dont la trajectoire est $G(A|E|P_o) = 0$ nous dirons, par analogie avec les résultats du paragraphe précédent, que $t'-t'_o$ et r' sont les coordonnées de P par rapport à $0'(P_o$, E), et en déterminant ces fonctions en termes de P, E et P_o nous obtiendrons les formules de transformation de coordonnées entre C , observateur fixe à r = 0, et la classe d'observateurs en chute libre 0'.

4. - Détermination de la fonction $t'-t'_o$.

La résolution de la première partie du problème est presque immédiate. En effet la fonction $\tau(A|E|P_o)$, pour définir une famille de transversales, doit être une intégrale complète de l'équation de Hamilton-Jacobi associée à (IV, 14a) qui est

$$\frac{1}{\sigma}\left(\frac{\partial \tau}{\partial t}\right)^2 - c^2\sigma\left(\frac{\partial \tau}{\partial r}\right)^2 = 1 \quad .$$

D'autre part, la constante d'intégration devant être l'Hamiltonien, $\tau(A|E|P_o)$ doit être de la forme :

$$\tau(A|E|P_o) = -E(t-t_o) + W(r, r_o, E)$$

d'où en obtient facilement

$$\text{(IV, 17)} \quad \tau(A|E|P_o) = -E(t-t_o) - \frac{\varepsilon}{c}\int_{r_o}^{r} \frac{\Omega}{\sigma}\, dr \qquad \Omega = \sqrt{E^2 - \sigma} \quad , \quad \varepsilon = \frac{v}{|v|} \;.$$

Si P_1 est le point d'intersection de la variété $\tau(A|E|P_o) = \tau(P|E|P_o)$ avec l'extrémale, $G(A|E|P_o) = 0$, pour être sur la variété nous aurons

$$\tau(P_1|E|P_o) = \tau(P|E|P_o)$$

et pour être sur l'extrémale d'après le théorème du Chap. I, Sect. I :

$$\tau(P_1|E|P_o) = \tau_{P_o}^{P_1}$$

et par conséquent nous obtenons finalement

$$\text{(1V, 18)} \qquad t'-t'_o = -E(t-t_o) - \frac{\epsilon}{c}\int_{r_o}^{r} \frac{\Omega}{\sigma}\, dr$$

La méthode de Hamilton-Jacobi permet ainsi de résoudre facilement un problème d'apparence compliqué et qui le serait même réellement si on essayant d'appliquer directement la définition de la fonction $t'-t'_o$.

5. - <u>Détermination de la fonction r'.</u>

a. - Résolvons maintenant la deuxième partie. Començons par constater que la méthode de Hamilton-Jacobi n'est pas dans ce cas directement applicable. En effet, nous pourrions appliquer cette méthode si la variété $\tau(A|E|P_o) = \tau(P|E|P_o)$ était une extrémale de

$$\text{(IV, 19)} \qquad \mathcal{H}_{P_1}^{P} = \int_{t_1}^{t} \epsilon\sqrt{\frac{u^2}{\sigma} - c^2\sigma}\; dt \equiv \int_{t_1}^{t} L\, dt$$

et si $G(A|E|P_o) = 0$ était une variété transversale aux extrémales $\tau(A|E|P_o)$ = cte. Or il n'en est rien. En effet, les équations différentielles des extrémales de (IV, 19) sont $\mathcal{E} = 0$, où :

$$\mathcal{E} \equiv \frac{d}{dt} \frac{\partial L}{\partial u} - \frac{\partial L}{\partial r} .$$

Or :

$$\frac{\partial L}{\partial u} = \frac{\varepsilon u}{\sigma \sqrt{\frac{u^2}{\sigma} - c^2 \sigma}} , \quad \frac{\partial L}{\partial r} = \frac{-\varepsilon (\frac{u^2}{\sigma^2} + c^2) \sigma'}{2\sqrt{\frac{u^2}{\sigma} - c^2 \sigma}} , \quad \sigma' = \frac{d\sigma}{dr} = \frac{2\mu}{r^2}$$

et puisque sur $\tau(A|E|P_o)$ = cte, d'après (IV, 17), nous avons:

$$u = \frac{-\varepsilon c E \sigma}{\Omega} \tag{IV, 20}$$

par un calcul facile nous obtenons :

$$\frac{\partial L}{\partial u} = -\frac{E}{\sigma} \qquad \frac{\partial L}{\partial r} = \frac{-\varepsilon c(2E^2 - \sigma)}{2\sigma\Omega} \sigma' . \tag{IV, 21}$$

Ensuite, de :

$$\frac{d}{dt} \frac{\partial L}{\partial u} = \frac{E}{\sigma^2} \sigma' u = -\frac{\varepsilon c E^2}{\sigma \Omega} \sigma'$$

il vient finalement

$$\mathcal{E} = -\frac{\varepsilon c \sigma'}{2\Omega} \neq 0 \tag{IV, 22}$$

ce qui prouve que les variétés $\tau(A|E|P_o)$ = cte ne sont pas des extrémales de (IV, 19).

b. - Pour résoudre ces difficultés nous introduisons l'intégrale :

$$\int_{t_1}^{t} \Lambda \, dt \equiv \int_{t_1}^{t} \left[L + \lambda (E + \frac{\varepsilon}{c} \frac{\Omega}{\sigma} u) \right] dt \tag{IV, 23}$$

où $\lambda(r, t, E)$ est une fonction auxiliaire. D'après (IV, 20), le long de

$\tau(A|E|P_o)$ = cte :

$$E + \frac{\varepsilon}{c}\frac{\Omega}{\sigma}u = 0$$

et par conséquent la fonction r'(IV, 16b) peut tout aussi bien être définie par

(IV, 24) $$r' = \mathcal{H}_{P_1}^{P} = \int_{t_1}^{t} \Lambda \, dt$$

l'intégrale étant calculée le long de $\tau(A|E|P_o) = \tau(P|E|P_o)$.

Imposons à λ les deux conditions suivantes :

1) λ est telle que les variétés $\tau(A|E|P_o)$ = cte sont des extrémales de (IV, 23).

2) l'extrémale $G(A|E|P_o) = 0$ de (IV, 14a) est une variété transversale à la famille d'extrémales $\tau(A|E|P_o)$ = cte.

Si une telle fonction λ existe nous pouvons appliquer la méthode de Hamilton-Jacobi pour calculer la fonction r' .

c. - Les variétés $\tau(A|E|P_o)$ = cte seront des extrémales de (IV, 23) si le long de ces variétés :

$$\bar{\mathcal{E}} \equiv \frac{d}{dt}\frac{\partial \Lambda}{\partial u} - \frac{\partial \Lambda}{\partial r} = \mathcal{E} + \frac{\varepsilon}{c}\frac{d}{dt}\left(\frac{\lambda\Omega}{\sigma}\right) - E\frac{\partial \lambda}{\partial r} - \frac{\varepsilon}{c}\frac{\partial}{\partial r}\left(\frac{\lambda\Omega}{\sigma}\right)u = 0 \quad .$$

Or :

$$\frac{d}{dt}\left(\frac{\lambda\Omega}{\sigma}\right) = \frac{\Omega}{\sigma}\frac{d\lambda}{dt} + \lambda\frac{\partial}{\partial r}\left(\frac{\Omega}{\sigma}\right)u$$

d'où il vient :

$$\bar{\mathcal{E}} = \mathcal{E} + \frac{\varepsilon\Omega}{c\sigma}\frac{d\lambda}{dt} - E\frac{\partial \lambda}{\partial r} - \frac{\varepsilon u\Omega}{c\sigma}\frac{\partial \lambda}{\partial r} = 0$$

qui compte tenu de (IV, 20) devient :

$$\bar{\mathcal{E}} = \mathcal{E} + \frac{\varepsilon\Omega}{c\sigma}\frac{d\lambda}{dt} = 0 \quad .$$

La condition 1) imposée à λ est donc, compte tenu de (IV, 22) :

(IV, 25)
$$\frac{d\lambda}{dt} \equiv \frac{\partial \lambda}{\partial t} + \frac{\partial \lambda}{\partial r} u = \frac{c^2 \sigma \sigma'}{2\Omega^2}$$

avec u donné par (IV, 20). Cette équation détermine λ sur chaque variété $\tau(A|E|P_o)$ = cte si on se donne sur chacune de ces variétés la valeur de λ en un point. Nous allons voir que c'est précisément la condition 2) qui fixe ces conditions initiales.

d. - Une fonction $\mathcal{H}(A|E)$ = cte sera une variété transversale à la famille d'extrémales $\tau(A|E|P_o)$ = cte de (IV, 23) si et seulement si (Voir Déf. Chap. I, Sect. I):

(IV, 26)
$$\frac{\partial \mathcal{H}}{\partial r} = \frac{\partial \Lambda}{\partial u} = \frac{\partial L}{\partial u} + \frac{\varepsilon}{c} \lambda \frac{\Omega}{\sigma}$$
$$\frac{\partial \mathcal{H}}{\partial t} = \Lambda - \frac{\partial \Lambda}{\partial u} = L + \lambda E - \frac{\partial L}{\partial u} u \quad .$$

Mais compte tenu de (IV, 20) :

$$L = \frac{\varepsilon c \sigma}{\Omega}$$

et d'après (IV, 21) nous pouvons écrire les conditions (IV, 26) sous la forme :

(IV, 27)
$$\frac{\partial \mathcal{H}}{\partial r} = -\frac{E}{\sigma} + \frac{\varepsilon}{c} \lambda \frac{\Omega}{\sigma}$$
$$\frac{\partial \mathcal{H}}{\partial t} = -\varepsilon c \Omega + \lambda E \quad .$$

Ceci nous donne un système d'équations en dérivées partielles pour déterminer $\mathcal{H}$ si nous connaissons λ. Car si λ satisfait la condition 1), donc (IV, 25), ce système d'équations est complètement intégrable. En effet :

(IV, 28)
$$\frac{\partial^2 \mathcal{H}}{\partial t \partial r} - \frac{\partial^2 \mathcal{H}}{\partial r \partial t} = \frac{\varepsilon}{c} \frac{\Omega}{\sigma} \frac{\partial \lambda}{\partial t} + \varepsilon c \frac{\partial \Omega}{\partial r} - E \frac{\partial \lambda}{\partial r} \quad .$$

Mais d'après (IV, 17) :

(IV, 29) $$\frac{\partial \Omega}{\partial r} = - \frac{\varepsilon \sigma'}{2\Omega}$$

et d'après (IV, 25) et (IV, 20) :

$$\frac{\partial \lambda}{\partial t} = \frac{c^2 \sigma \sigma'}{2\Omega^2} + \frac{\varepsilon c E \sigma}{\Omega} \frac{\partial \lambda}{\partial r}$$

d'où, par substitution de cette dernière et de (IV, 29) dans (IV, 28) on obtient:

$$\frac{\partial^2 \mathcal{H}}{\partial t \partial r} - \frac{\partial^2 \mathcal{H}}{\partial r \partial t} = 0 \quad .$$

Ce qu'il fallait démontrer .

e. - Supposons λ connue et soit $\mathcal{H}(A|E)$ = cte l'intégrale générale de (IV, 27). Calculons sa dérivée totale par rapport au temps le long de $G(A|E|P_o)=0$. De (IV, 15) il vient :

(IV, 30) $$v = - \frac{\varepsilon c \sigma \Omega}{E}$$

ce qui avec (IV, 27) conduit au résultat suivant :

(IV, 31) $$\frac{d\mathcal{H}}{dt} = \frac{\partial \mathcal{H}}{\partial t} + \frac{\partial \mathcal{H}}{\partial r} v = \frac{\lambda \sigma}{E} \quad .$$

Posons :

$$\mathcal{H}(A|E; P_o) = \mathcal{H}(A|E) - \mathcal{H}(P_o|E) \; .$$

Avec cette notation l'équation de la variété transversale de la famille $\mathcal{H}(A|E)$ = cte qui passe par P_o est :

$$\mathcal{H}(A|E; P_o) = 0$$

de sorte que en particulier nous avons :

$$\mathcal{H}(P_o|E; P_o) = 0 \quad .$$

Ainsi, la condition nécessaire et suffisante pour que l'extrémale $G(A|E|P_o)=0$ de (IV, 14a) coincide avec $\mathcal{H}(A|E; P_o) = 0$, et la condition 2) soit satisfaite, est que :

$$\frac{d\mathcal{H}}{dt}(A|E;\ P_o) = 0 \tag{IV, 32}$$

le long de $G(A|E;\ P_o) = 0$. Autrement dit, d'après (IV, 31), que λ soit zéro sur $G(A|E|P_o) = 0$. Cette condition avec (IV, 25) détermine complètement la fonction λ et par conséquent aussi l'intégrale générale de (IV, 26) $\mathcal{H}(A|E) =$ = cte.

f. - Considérons maintenant, P étant un évènement quelconque, la fonction :

$$\mathcal{H}(P|E;\ P_o) = \mathcal{H}(P|E) - \mathcal{H}(P_o,\ E)$$

et soit P_1 le point d'intersection de l'extrémale $\mathcal{T}(A|E|P_o) = \mathcal{T}(P|E|P_o)$ avec la variété transversale :

$$\mathcal{H}(A|E;P_o) = 0 \Longleftrightarrow G(A|E|P_o) = 0 \quad .$$

P_1 étant sur la même transversale que P_o, d'après (IV, 32), nous avons :

$$\mathcal{H}(P|E;P_o) = \mathcal{H}(P|E) - \mathcal{H}(P_o|E) = \mathcal{H}(P|E) - \mathcal{H}(P_1|E) = \mathcal{H}(P|E;P_1)$$

et P_1 et P étant sur la même extrémale, d'après le théorème du Chap. I, Sect. I :

$$\mathcal{H}(P|E;P_o) = \mathcal{H}(P|E;P_1) = \int_{P_1}^{P} \Lambda \ dt \quad .$$

Soit finalement, compte tenu de (IV, 24) :

$$r' = \mathcal{H}_{P_1}^{P} = \mathcal{H}(P|E;P_o)$$

et la deuxième partie du problème est résolue.

6. - Calcul explicite approché de la fonction r'.

Nous nous proposons de calculer explicitement la fonction r' jusqu'au deuxième ordre d'approximation. A cet ordre nous avons :

$$r' = \mathcal{H}(P|E;P_o) = \mathcal{H}_{P=P_o} + \frac{\partial \mathcal{H}}{\partial t}\Big|_{P=P_o}(t-t_o) + \frac{\partial \mathcal{H}}{\partial r}\Big|_{P=P_o}(r-r_o) +$$

$$+ \frac{1}{2}\frac{\partial^2 \mathcal{H}}{\partial t^2}\Big|_{P=P_o}(t-t_o)^2 + \frac{\partial^2 \mathcal{H}}{\partial r \partial t}\Big|_{P=P_o}(t-t_o)(r-r_o) + \frac{1}{2}\frac{\partial^2 \mathcal{H}}{\partial r^2}\Big|_{P=P_o}(r-r_o)^2 .$$

Or manifestement :

$$\mathcal{H}_{P=P_o} = 0$$

et de (IV, 27) il vient, puisque $\lambda_{P=P_o} = 0$:

$$\frac{\partial \mathcal{H}}{\partial r}\Big|_{P=P_o} = -\frac{E}{\sigma} \qquad \frac{\partial \mathcal{H}}{\partial t}\Big|_{P=P_o} = -\varepsilon cP ,$$

les deuxièmes membres étant calculés bien entendu au point P_o. Nous supprimerons l'indication $P=P_o$ quand il n'y aura pas de confusion à craindre

De (IV, 27) il vient

$$\frac{\partial^2 \mathcal{H}}{\partial r^2}\Big|_{P=P_o} = \frac{E}{\sigma^2}\sigma' + \frac{\varepsilon}{c}\frac{\partial \lambda}{\partial r}\Big|_{P=P_o} \cdot \frac{\Omega}{\sigma}$$

$$\text{(IV, 33)} \qquad \frac{\partial^2 \mathcal{H}}{\partial t \partial r}\Big|_{P=P_o} = \frac{\varepsilon}{c}\frac{\partial \lambda}{\partial t}\Big|_{P=P_o}\frac{\Omega}{\sigma}$$

$$\frac{\partial^2 \mathcal{H}}{\partial t^2}\Big|_{P=P_o} = \frac{\partial \lambda}{\partial t}\Big|_{P=P_o} E .$$

Nous avons donc besoin de connaître $\frac{\partial \lambda}{\partial r}\Big|_{P=P_o}$ et $\frac{\partial \lambda}{\partial t}\Big|_{P=P_o}$. Or de (IV, 25) il vient

$$\text{(IV, 34)} \qquad \frac{\partial \lambda}{\partial t}\Big|_{P=P_o} - \frac{\varepsilon cE\sigma}{\Omega}\frac{\partial \lambda}{\partial r}\Big|_{P=P_o} = \frac{c^2\sigma\sigma'}{2\Omega^2}$$

D'autre part, puisque λ est zéro sur $G(A|E|P_o) = 0$ nous aurons le long de $G(A|E|P_o) = 0$:

$$\frac{\partial\lambda}{\partial t} + \frac{\partial\lambda}{\partial r} v = 0$$

ce qui compte tenu de (IV, 30) nous donne en particulier au point P_o :

$$\left.\frac{\partial\lambda}{\partial t}\right|_{P=P_o} - \frac{\varepsilon c \sigma \Omega}{E} \left.\frac{\partial\lambda}{\partial r}\right|_{P=P_o} = 0 . \tag{IV, 35}$$

De (IV, 34) et (IV, 35) on tire les résultats suivants :

$$\left.\frac{\partial\lambda}{\partial t}\right|_{P=P_o} = -\frac{c^2\sigma'}{2} \qquad \left.\frac{\partial\lambda}{\partial r}\right|_{P=P_o} = -\frac{\varepsilon c E \sigma'}{2\sigma\Omega} .$$

La substitution de ces résultats dans (IV, 33) donne :

$$\left.\frac{\partial^2\mathcal{H}}{\partial r^2}\right|_{P=P_o} = \frac{E\sigma'}{2\sigma^2} \qquad \left.\frac{\partial^2\mathcal{H}}{\partial t\partial r}\right|_{P=P_o} = -\frac{\varepsilon c\Omega\sigma'}{2\sigma} \qquad \left.\frac{\partial^2\mathcal{H}}{\partial t^2}\right|_{P=P_o} = -\frac{c^2\sigma' E}{2} .$$

Et nous pouvons finalement écrire la fonction r', à l'approximation envisagée, sous la forme :

$$\begin{aligned} r' = \mathcal{H}(P|E; P_o) = &- \varepsilon c \Omega (t-t_o) - \frac{E}{\sigma}(r-r_o) \\ &- \frac{1}{4} c^2 E \sigma'(t-t_o)^2 - \frac{1}{2}\frac{\varepsilon c \Omega}{\sigma}\sigma'(t-t_o)(r-r_o) + \frac{E\sigma'}{2\sigma^2}(r-r_o)^2 . \end{aligned} \tag{IV, 36}$$

7. - Formes limite des fonctions de transformation.

a. - Que les transformations (IV, 18) et (IV, 36) se réduisent aux transformations (IV, 6) quand kM = 0, résulte des définitions (IV, 16) et des résultats (IV, 12) que nous avons prouvés au sujet du groupe de Lorentz.

b. - Démontrons qu'elles se réduisent à (IV, 1) et (IV, 3) quand on fait tendre c vers infini. De (IV, 13), (IV, 15) et (IV, 17) il vient :

$$\lim_{c\to\infty} \sigma = 1 \qquad \lim_{c\to\infty} E = -1 \qquad \lim_{c\to\infty} \Omega = 0 \tag{IV, 37}$$

Par conséquent pour c tendant vers infini (IV, 17) devient :

$$t'-t'_o = t-t_o \quad .$$

Pour la transformation d'espace les choses sont plus délicates. Nous esquissons ici seulement la démonstration. Nous pouvons écrire (IV, 19) sous la forme

$$r' = \int_{t_1}^{t} \varepsilon \sqrt{\frac{u^2}{\sigma} - c^2\sigma} \, dt = \int_{r_1}^{r} \sqrt{\frac{1}{\sigma} - \frac{c^2\sigma}{u^2}} \, dr \quad .$$

Mais d'après (IV, 20) :

$$\frac{c^2\sigma}{u^2} = \frac{\Omega^2}{E^2\sigma}$$

et par conséquent de (IV, 37) :

$$\lim_{c \to \infty} \frac{c^2\sigma}{u^2} = 0$$

et

(IV, 38)
$$r' = \int_{r_1}^{r} dr = r-r_1$$

qui est (IV, 2). Que les arguments de (IV, 38) et (IV, 2) sont les mêmes résulte du fait bien connu que pour $c \to \infty$ les équations du mouvement dans le cas du ds^2 de Schwarzschild se réduisent à celles de la théorie de Newton.

Remarque. - Il convient de remarquer que les résultats de ce chapitre sont purement locaux, c'est-à-dire supposent P suffisamment voisin de P_o. Ceci venant essentiellement du fait que ε peut changer de signe. Ceci pose des ambiguités que nous ne discuterons pas ici. Signalons tout-de-même que si P n'est pas suffisamment voisin de P_o le calcul des fonctions $t'-t'_o$ et r' doit se faire en deux étapes successives.

SECTION II :

PARTIE PRINCIPALE DE LA GENERALISATION DU GROUPE DE LORENTZ INHOMOGENE

1. - Le groupe de Lorentz inhomogène.

Soit 0' un observateur galiléen qui à l'instant t_o occupe la position x^i_o par rapport à un second observateur galiléen 0 , et a par rapport à lui une vitesse dont les composantes sont v^i. Supposons 0 et 0' munis de tri-repères orthonormés S et S' et soit $R^{i'}_j$ la matrice du groupe orthogonal qui définit l'orientation relative du tri-repère S' par rapport à S . Symboliquement S' = RS . Dans ces conditions les formules de transformation des coordonnées entre 0 et 0' sont les suivantes :

(IV, 39)
$$\theta^{o'} = - P_o \theta^o + P_i \theta^i$$
$$\theta^{i'} = \sum_j R^{i'}_j \left\{ \left[\delta_{jk} + \frac{c^2 P_j P_k}{1 - P_o} \right] \theta^k - c^2 P_j \theta^o \right\}$$

où :

$$P_o = \frac{-1}{\sqrt{1 - \frac{v^2}{c^2}}} \qquad P_i = \frac{v^i}{c^2} P_o$$

(IV, 40)
$$\theta^{o'} = t' - t'_o \;, \quad \theta^{i'} = x^{i'} \;; \quad \theta^o = t - t_o \;; \quad \theta^i = x^i - x^i_o \;.$$

L'interprétation cinématique de ces formules est la suivante : si (x^i, t) sont les coordonnées d'un évènement quelconque A par rapport à 0 , $(x^{i'}, t')$ sont les coordonnées du même évènement A par rapport à 0'. t'_o est le temps de 0' qui correspond à t_o.

Les formules (IV, 39) sont les formules de transformation définissant le groupe de Lorentz inhomogène.

D'après les définitions des P_o et P_i il vient :

$$(IV, 41) \qquad P_o^2 - c^2 \sum P_i^2 = 1$$

ce qui entraîne, comme on sait, quelle que soit $R_j^{i'}$ du groupe orthogonal :

$$(\theta^{o'})^2 - \frac{1}{c^2} \sum_{i'} (\theta^{i'})^2 = (\theta^o)^2 - \frac{1}{c^2} \sum_i (\theta^i)^2 .$$

Encore convient-il d'insister sur le fait, qui nous servira plus tard, que le résultat précédent est purement algébrique. Il ne dépend que de (IV, 41) et nullement de l'interprétation des θ qui peut être celle de (IV, 40) ou toute autre.

Par différentiation les formules (IV, 39) restent manifestement inchangées à condition de poser :

$$(IV, 42) \qquad \theta^{o'} = dt' \; , \quad \theta^{i'} = dx^{i'} \; , \quad \theta^o = dt \; , \quad \theta^i = dx^i \; .$$

Considérons les ds^2 de Minkowski :

$$ds^2 = dt^2 - \frac{1}{c^2} \sum (dx^i)^2 \quad .$$

dt et dx^i définissent en chaque point un corepère orthonormé. Les formules (IV, 39) avec (IV, 42) et compte tenu de (IV, 41) définissent sur ce corepère une transformation du groupe de Lorentz homogène. Réciproquement, à partir des formules (IV, 39) et l'interprétation de θ de (IV, 42) on obtient trivialement les formules (IV, 39) avec l'interprétation des θ donnée par (IV, 40), par intégration de quatre différentielles exactes.

2. - Partie principale de la généralisation du groupe de Lorentz inhomogène.

a. - Soit C un observateur fixe à $r = 0$ et soit 0' un observateur en chute libre décrivant une extrémale de (I, 16) ou, ce qui revient au même, une extrémale de

$$\tau_{P_o}^{P_1} = \int_{t_o}^{t} \Sigma \, dt \quad , \quad \Sigma = \sqrt{\sigma - \frac{\dot{r}^2}{c^2\sigma} - \frac{r^2}{c^2}(\dot{\theta}^2 + \sin^2\theta\,\dot{\varphi}^2)} \tag{IV, 42}$$

qui est l'intégrale de temps propre. L'observateur 0' sera caractérisé par sa position initiale $(r_o, \theta_o, \varphi_o)$ pour $t = t_o$ et pour des valeurs fixes des trois intégrales premières E , P et M définies au Chap. I, Sect. II, à cette diffèrence près qu'il faut poser $m = -\frac{1}{c^2}$ pour se conformer à (IV, 42). Il faudra d'ailleurs en faire de même dans la suite de cette section chaque fois que nous nous référerons à un résultat précédent pour lequel nous avions supposé m arbitraire. Nous écrirons symboliquement $0'(A_o, E, P, M)$.

Nous utiliserons par la suite des coordonnées polaires pour nous reférer à C et des coordonnées rectilignes pour nous référer à 0' . Ceci s'impose par le fait que, à cause de la symétrie sphérique du champ considéré, les coordonnées polaires sont plus commodes pour décrire certains aspects du problème que nous traitons ici, et que, à cause de la simplicité des formules de transformation du groupe de Lorentz en coordonnées rectilignes, ces dernières sont plus commodes pour d'autres aspects.

Soit A un évènement de coordonnées (ξ^i, t) par rapport à C . Les coordonnées de ce même évènement par rapport à $0'(A_o, E, P, M)$ seront données par des formules de la forme

$$t'-t'_o = \tau(A|E, P, M|A_o) \qquad x^{i'} = \mathcal{H}^i(A|E, P, M|A_o) \tag{IV, 43}$$

t'_o étant le temps pour 0' qui correspond à t_o . Evidemment nous devons avoir :

$$\tau(A_o|E, P, M|A_o) = 0 \qquad \mathcal{H}^i(A_o|E, P, M|A_o) = 0 \quad .$$

Si l'évènement A est suffisamment proche de A_o nous pourrons écrire au premier ordre d'approximation :

$$t'-t'_o = \frac{\partial \mathcal{T}}{\partial t}\bigg|_{A=A_o}(t-t_o) + \sum_j{}' \frac{\partial \mathcal{T}}{\partial \xi^j}\bigg|_{A=A_o}(\xi^j - \xi^j_o)$$

$$x^{i'} = \frac{\partial \mathcal{H}^i}{\partial t}\bigg|_{A=A_o}(t-t_o) + \sum_j{}' \frac{\partial \mathcal{H}^i}{\partial \xi^j}\bigg|_{A=A_o}(\xi^j - \xi^j_o)$$

formules qui sont la partie principale des transformations (IV, 43). D'une manière équivalente, nous pouvons nous référer à cette partie principale en posant

$$dt' = d\mathcal{T} \qquad\qquad dx^{i'} = d\mathcal{H}^i$$

étant entendu qu'il s'agit de formules pour $A=A_o$ et que par conséquent ces formules n'expriment pas que dt' et $dx^{i'}$ sont des différentielles exactes de fonctions des ξ^i et t.

Notre but dans cette section est d'imposer des conditions à dt' et $dx^{i'}$ et de les déterminer en identifiant chacune des grandeurs qui apparaisse dans les résultats.

b. - Nous désignerons dans la suite le point A par $A_o + \overrightarrow{dA}$.

Nous imposons à dt' les deux conditions suivantes

1) Si $\overrightarrow{dA}$ est tangent à l'extrémale $G^i(A|E,P,M|P_o) = 0$, trajectoire de 0', au point A_o :

$$dt' = \sum{}' dt$$

2) Si $\overrightarrow{dA}$ est une direction transversale à cette extrémale

$$dt' = d\mathcal{T} = 0 \quad .$$

Considérons l'intégrale complète (I, 23) de l'équation de Hamilton-Jacobi (I, 20), ou plutôt sa différentielle

$$F^o \equiv d\mathcal{T} = -E\,dt - \frac{\varepsilon}{c}\frac{\Omega}{\sigma}\,dr - \gamma\sqrt{P^2 - \frac{M^2}{\sin^2\theta}}\,d\theta - M\,d\varphi$$

Nous avons écrit τ au lieu de S pour rappeler que $m = -\frac{1}{c^2}$. Cependant nous avons déjà tenu compte de cette modification en ce qui concerne les signes de sorte que nous continuons à avoir

$$\varepsilon = \frac{\dot{r}}{|\dot{r}|} \qquad \gamma = \frac{\dot{\theta}}{|\dot{\theta}|} \qquad \frac{M}{|M|} = \frac{\dot{\varphi}}{|\dot{\varphi}|} \quad .$$

Nous supposons :

$$\text{(IV, 44)} \qquad \Omega \neq 0 \qquad \sqrt{P^2 - \frac{M^2}{\sin^2\theta}} \neq 0$$

au point A_o. Les résultats qui suivent seront valables indépendamment de ces restrictions mais la méthode que nous allons exposer, dans le cas où nous n'aurions pas (IV, 44), devrait être convenablement adaptée.

On obtiendrait les équations de la trajectoire de $0'(A_o, E, P, M)$ par la méthode habituelle rappelée au Chap. I. Ces équations, sous forme différentielle, sont :

$$F^i = 0$$

où :

$$F^1 \equiv -dt - \frac{\varepsilon}{c}\frac{E}{\sigma\Omega}dr$$

$$F^2 \equiv \frac{\varepsilon\, cP}{r^2\Omega}dr - \frac{\gamma\, Pd\theta}{\sqrt{P^2 - \frac{M^2}{\sin^2\theta}}}$$

$$F^3 \equiv \frac{\gamma\, M\, d\theta}{\sin^2\theta\sqrt{P^2 - \frac{M^2}{\sin^2\theta}}} - d\varphi$$

$\vec{dA}(dt, dr, d\theta, d\varphi)$ sera tangent à l'extrémale envisagée si

$$F^i(\vec{dA}) = 0 \quad .$$

D'autre part $\vec{dA}$ sera transversale à cette extrémale si

$$F^o(\vec{dA}) = 0 \quad .$$

Les quatre formes de Pfaff F^{α} ($\alpha = 0, 1, 2, 3$) sont linéairèment indépendantes. En effet si nous posons

$$F^{\alpha} = F^{\alpha}{}_{\beta}\, d\xi^{\beta}$$

nous avons :

$$\text{dét}\left| F^{\alpha}{}_{\beta} \right| = - \frac{\varepsilon \mathcal{J} P}{c\Omega \sqrt{P^2 - \frac{M^2}{\sin^2\theta}}} \neq 0 \quad \text{et fini}$$

d'après (IV, 44).

Par conséquent dt' sera une combinaison linéaire des F^{α}

$$dt' = \lambda_o F^o + \lambda_i F^i \quad .$$

Or d'après la condition 1) nous devons avoir

$$dt' = \Sigma\, dt \qquad \text{si } F^i = 0$$

mais d'après le théorème du Chap. I, Sect. I :

$$\Sigma\, dt = d\tau = F^o \qquad \text{si } F^i = 0$$

donc

$$\lambda_o = 1 \quad .$$

D'autre part d'après la condition 2) nous devons avoir

$$\text{(IV, 45)} \qquad dt' = \lambda_i F^i = 0 \qquad \text{si } d\tau = 0 \quad .$$

Si $d\tau = 0$ nous avons

$$- dt = \frac{\varepsilon}{c} \frac{\Omega}{E\sigma}\, dr + \frac{\mathcal{J}}{E} \sqrt{P^2 - \frac{M^2}{\sin^2\theta}}\, d\theta + \frac{M}{E} d\varphi \quad .$$

Nous pouvons par conséquent écrire (IV, 45) sous la forme :

$$\text{(IV, 46)} \qquad dt' = \lambda_i F^{i'} = 0$$

où

$$F^{1'} = \left(\frac{\varepsilon}{c}\frac{\Omega}{E\sigma} - \frac{\varepsilon}{c}\frac{E}{\sigma\Omega}\right) dr + \frac{\gamma}{E}\sqrt{P^2 - \frac{M^2}{\sin^2\theta}}\, d\theta + \frac{M}{E}\, d\varphi$$

$$F^{2'} = F^2 \qquad\qquad F^{3'} = F^3 .$$

Les formes de Pfaff $F^{i'}$ ne dépendent que de dr, $d\theta$, et $d\varphi$ qui sont arbitraires puisque nous avons déjà tenu compte de $d\tau = 0$, et elles sont linéairement indépendantes. En effet :

$$\text{dét}\left|F^{i'}_j\right| = \frac{-\varepsilon\gamma P}{cE\Omega\sqrt{P^2 - \frac{M^2}{\sin^2\theta}}}$$

(IV, 46) entraîne donc :

$$\lambda_i = 0$$

et par conséquent nous aurons quel que soit $\overrightarrow{dA}$

$$dt' = d\tau = -E\,dt - \frac{\varepsilon}{c}\frac{\Omega}{\sigma}\, dr - \gamma\sqrt{P^2 - \frac{M^2}{\sin^2\theta}}\, d\theta - M\, d\varphi .$$

Remarque. - L'expression ci-dessus est évidemment complètement intégrable. Cependant il serait tout-à-fait inexact de prétendre que l'intégrale générale de cette équation serait la généralisation de la transformation temporelle du groupe de Lorentz inhomogène. Ceci vient du fait que P et M ne sont pas les bonnes constantes d'intégration pour le problème. Mais tant qu'on reste à la partie principale le choix des constantes d'intégration est sans importance.

c. - En accord avec ce qu'on peut appeler le principe de validité locale de la Relativité Restreinte la condition pour déterminer les $dx^{i'}$ est la suivante : nous supposons que les $dx^{i'}$ sont tels que :

(IV, 47) $$ds^2 = (\omega^o)^2 - \frac{1}{c^2}\sum_i (\omega^i)^2 = dt'^2 - \frac{1}{c^2}\sum_{i'} (dx^{i'})^2$$

où nous avons posé

$$\omega^o = \sqrt{\sigma}\, dt \qquad \omega^1 = \frac{dr}{\sqrt{\sigma}} \qquad \omega^2 = r\, d\theta \qquad \omega^3 = r \sin\theta \; d\varphi \, .$$

Or si nous posons :

$$Q_o = \frac{E}{\sqrt{\sigma}} \qquad Q_1 = -\frac{\varepsilon}{c}\frac{\Omega}{\sqrt{\sigma}} \qquad Q_2 = -\frac{\gamma}{r}\sqrt{P^2 - \frac{M^2}{\sin^2\theta}} \qquad Q_3 = -\frac{M}{r \sin\theta}$$

nous avons :

$$dt' = - Q_o\, \omega^o + Q_i\, \omega^i \quad .$$

D'autre part, d'après la définition des Q_α il vient

$$Q_o^2 - c^2 \sum_i Q_i^2 = 1$$

et par conséquent d'après la remarque que nous avons faite au paragraphe 1 , la solution générale pour les $dx^{i'}$ satisfaisant à la condition (IV, 47) est

$$dx^{i'} = \sum_j R_j^{i'} \left\{ \left[\delta_{jk} + c^2 \frac{Q_j Q_k}{1 - Q_o} \right] \omega^k - c^2 Q_j\, \omega^o \right.$$

$R_j^{i'}$ étant une matrice quelconque du groupe orthogonal qui peut être interprétée comme définissant l'orientation relative du trirepère de 0' par rapport au tri-repère orthonormé adapté aux coordonnées polaires au point $A = A_o$.

APPENDICE A [1], [2]

Rappels sur les équations différentielles linéaires au deuxième ordre.

a. - Considérons l'équation différentielle linéaire du deuxième ordre :

$$\frac{d^2R}{dr^2} + M\frac{dR}{dr} + NR = 0 \tag{A, 1}$$

où M et N sont des fonctions de r. Par un changement de fonction inconnue toute équation de ce type peut être ramenée à une forme canonique où la dérivée première n'y figure pas. En effet, si nous posons :

$$R = F \exp\left[-\frac{1}{2}\int M dr\right]$$

la nouvelle équation pour F est :

$$\frac{d^2F}{dr^2} + JF = 0$$

où :

$$J = N - \frac{1}{2}\frac{dM}{dr} - \frac{1}{4}M^2$$

b. - Nous nous bornerons dans ces rappels au cas où M et N sont des fonctions rationnelles. Dans ce cas l'équation peut toujours s'écrire sous la forme :

$$P_o\frac{d^2R}{dr^2} + P_1\frac{dR}{dr} + P_2R = 0 \tag{A, 2}$$

où P_i $(i = 0, 1, 2)$ sont des polynomes en r sans diviseur commun.

Les points singuliers à distance finie de l'équation (A, 2) sont les racines de l'équation $P_o = 0$. Tout autre point est dit ordinaire. Au voisinage de tout point ordinaire $r = a$ il existe une et une seule solution telle que :

$$R(a) = a_o \qquad \frac{dR}{dr}(a) = a_1$$

a_o et a_1 étant deux constantes arbitraires.

Il nous intéresse surtout ici quelques définitions et résultats concernant les points singuliers. Nous supposerons que $r = 0$ est un point singulier. L'étude qui suit s'appliquera évidemment à tout autre point singulier $r = p$ puisque on se ramène trivialement au cas précédent par le changement de variable $x = r-p$.

Posons :

$$\omega_i = \nu(P_i) \qquad i = 0, 1, 2$$

où $\nu(P_i)$ désigne la valuation du polynome correspondant. ω_i est l'ordre de multiplicité de la racine $r = 0$ pour chacune des équations $P_i = 0$. $r = 0$ étant un point singulier nous avons :

$$\omega_o > 0$$

Soit α un entier tel que :

$$\alpha + \omega_o \geqslant 2 \qquad \alpha + \omega_1 \geqslant 1 \qquad \alpha + \omega_2 \geqslant 0$$

l'égalité étant atteinte au moins une fois. Récrivons l'équation (A, 2) sous la forme :

$$D(R) \equiv Q_o r^2 \frac{d^2R}{dr^2} + Q_1 r \frac{dR}{dr} + Q_2 R = 0$$

où nous avons posé :

$$Q_o = r^{\alpha-2} P_o \qquad Q_1 = r^{\alpha-1} P_1 \qquad Q_2 = r^{\alpha} P_2$$

Nous avons :

$$\nu(Q_o) = \alpha - 2 + \omega_o > 0 \qquad \nu(Q_1) = \alpha - 1 + \omega_1 \geqslant 0 \qquad \nu(Q_2) = \alpha + \omega_2 \geqslant 0$$

l'égalité précédent le zéro étant atteinte au moins une fois.

Ecrivons les polynomes Q_i sous la forme :

(A, 3) $$Q_i = \sum_{o}^{\nu} Q_{i,h} r^h$$

où :

$$\nu = \max \text{dég} Q_i$$

D'après les définitions précédentes nous savons que

$$Q_{i,o} \neq 0$$

pour au moins une valeur de i , ainsi que

$$Q_{j,\nu} \neq 0$$

pour au moins une valeur de j .

c. - Nous nous proposons de savoir sous quelles conditions l'équation (A, 2) admet au voisinage de $r = 0$ des solutions dont le développement en série dans ce voisinage est de la forme

(A, 4) $$R = \sum_{n=o}^{\infty} a_n r^{s+n} \quad ; \quad a_o \neq 0$$

s étant une constante. De telles solutions si elles existent sont dites régulières. Elles ne sont pas les seules à porter ce nom, ce pourquoi nous les appellerons plus précisément solutions régulières au sens stricte.

Calculons $D(r^m)$. Nous obtenons facilement :

(A, 5) $$D(r^m) = r^m \left[m(m-1)Q_o + mQ_1 + Q_2 \right]$$

Pour que le développement présumé de R soit le développement d'une solution de $D(R)$ il faut que :

$$D(R) = \sum_{n=o}^{\infty} a_n D(r^{s+n}) = 0$$

ce que nous pouvons écrire, d'après (A, 5) et (A, 3), sous la forme :

$$D(R) = \sum_{n=o}^{\infty} a_n r^{s+n} \left[(s+n)(s+n-1) \sum_{k=o}^{\nu} Q_{o,k} r^k + (s+n) \sum_{k=o}^{\nu} Q_{1,k} r^k + \sum_{k=o}^{\nu} Q_{2,k} r^k = 0 \right]$$

ou encore

$$D(R) = \sum_{p=s}^{\infty} c_p r^p = 0 \tag{A, 6}$$

où

$$c_p = \sum_{n=o}^{p-s} a_n \left[(s+n)(s+n-1) Q_{o,p-s-n} + (s+n) Q_{1,p-s-n} + Q_{2,p-s-n} \right]$$

avec $p \geqslant s$. (A, 6) entraîne :

$$c_p = 0 \qquad\qquad p \geqslant s . \tag{A, 7}$$

En particulier pour $p = s$ on obtient

$$c_s = a_o \left[s(s-1) Q_{o,o} + s Q_{1,o} + Q_{2,o} \right] = 0$$

et puisque $a_o \neq 0$:

$$s(s-1) Q_{o,o} + s Q_{1,o} + Q_{2,o} = 0 \tag{A, 8}$$

Cette équation est dite l'équation indicielle. Elle exprime une condition nécessaire que s doit satisfaire pour que (A, 4) puisse être une solution de (A, 2).

d. - Nous nous proposons maintenant de caractériser la nature du point singulier par une étude de l'équation indicielle et rappeler quelques résultats sur l'existence de solutions régulières.

1) Si $Q_{o,o} \neq 0$ l'équation indicielle admet deux racines distinctes ou une racine double. Le point singulier est dit être un point singulier régulier. En termes des ω_i ce cas est caractérisé par :

$$\omega_o \leq \omega_1 + 1 \qquad\qquad \omega_o \leq \omega_2 + 2$$

Soient s_1 et s_2 les deux racines de (A, 8). Deux cas sont à distinguer :

a) Si s_1-s_2 n'est pas un entier il existe pour chacune des racines s_1 et s_2 une solution régulière au sens stricte de la forme (A, 4).

b) Si $s_1-s_2 > 0$ est un entier il n'existe en général qu'une seule solution régulière au sens stricte. Elle est associée à s_1 et on la détermine comme dans le cas précédent.

Si $s_1=s_2$ il ne peut exister évidemment qu'une seule solution régulière au sens stricte et elle existe réellement.

2) Si $Q_{o,o} = 0$, $Q_{1,o} \neq 0$, ce qui entraîne

$$\omega_o > \omega_1 + 1 \qquad \omega_1 \leq \omega_2 + 1$$

l'équation indicielle n'admet qu'une seule racine. Nous dirons dans ce cas que $r = 0$ est un point quasi-régulier. La solution régulière associée à cette racine peut exister ou peut ne pas exister.

Dans tous les cas où la solution régulière au sens stricte peut exister les coefficients a_n se calculent à partir des équations (A, 7) pour $p > s$ qui constituent une relation de récurrence pour ces coefficients contenant au maximum, comme on peut le voir facilement, $\nu +1$ coefficients. Le développement ainsi obtenu converge dans les cas 1) pour s_1 et s_2 si s_1-s_2 n'est pas un entier, converge pour s_1 si s_1-s_2 est un entier positif et converge aussi pour $s_1=s_2$. Dans le cas 2) la solution régulière existe ou pas suivant que le développement obtenu converge ou pas. Ils existent des exemples pour lesquels l'une ou l'autre de ces circonstances se produit. Dans tous les cas où le développement obtenu est convergent le domaine de convergence n'est en général que le cercle de centre $r=0$ et rayon p, p étant le module du point singulier le plus proche de $r = 0$.

3) Enfin si $Q_{o,o} = 0$, $Q_{1,o} = 0$ alors nécessairement $Q_{2,o} \neq 0$ et l'équation indicielle n'a pas de solutions et par conséquent l'équation (A, 2) n'admet pas de solutions régulières au voisinage de $r = 0$. Le point $r = 0$ est dit

être un point singulier irrégulier. On peut caractériser aussi ce cas par

$$\omega_o > \omega_2 + 2 \qquad \omega_1 > \omega_2 + 1$$

e. - Pour connaître la nature de l'infini $r = \infty$ il suffit de faire le changement de variable $r = \frac{1}{u}$. La nature de $r = \frac{1}{u}$ est alors la nature du point $u = 0$ pour l'équation transformée. Tout ce que nous venons de dire s'applique donc après ce changement de variable. Il peut toutefois être plus commode d'avoir la caractérisation du point de l'infini d'une manière directe. Pour cela il n'y a qu'à suivre pas à pas ce que nous avons fait pour le point $r = 0$ en partant cependant cette fois de la définition suivante :

Une solution de (A, 2) est dite régulière au sens stricte au voisinage de l'infini si elle est de la forme

$$R = \sum_{n=o}^{\infty} b_n r^{\rho - n} \qquad b_o \neq 0$$

où ρ est une constante.

L'équation indicielle pour ρ est :

$$\rho(\rho-1)Q_{o,\nu} + \rho Q_{1,\nu} + Q_{2,\nu} = 0 \tag{A, 9}$$

1) Si $Q_{o,\nu} \neq 0$, le point de l'infini est soit un point ordinaire, soit un point singulier régulier. Il est ordinaire si et seulement si :

$$\text{dèg } P_o = \text{dèg}(2rP_o - r^2P_1) = \text{dèg } r^4P_2$$

2) Si $Q_{o,\nu} = 0$ et $Q_{1,\nu} \neq 0$, $r = \infty$ est un point singulier quasi-régulier.

3) Si $Q_{o,\nu} = Q_{1,\nu} = 0$, nécessairèment $Q_{2,\nu} \neq 0$, Le point $r = \infty$ est dans ce cas un point singulier irrégulier.

Au sujet de l'existence de solutions régulières au sens stricte les mêmes résultats que nous avons rappelés pour le point $r = 0$ s'appliquent ici avec une seule modification. Quand l'équation indicielle admet deux racines ρ_1 et ρ_2 qui diffèrent par un entier $\rho_1 - \rho_2 > 0$ seule l'existence de la solution régulière au sens stricte associée à ρ_2 est assurée.

Dans tous les cas où la solution régulière au sens stricte existe, ou peut exister, la suite d'équations qui permet de calculer les coefficients b_n de proche en proche est :

(A, 10) $$c_p = 0 \qquad p \geqslant -\rho - \nu + 1$$

où :

$$c_p \equiv \sum_{n=0}^{p+\rho+\nu} b_n \left[(\rho - n)(\rho - n - 1) Q_{o, n-\rho-p} + (\rho - n) Q_{1, n-\rho-p} + Q_{2, n-\rho-p} \right]$$

ou encore en posant $n = m + \tau$, $p + q = m$

$$c_{m-\rho} = \sum_{\tau=0}^{\nu} b_{m+\tau} \left[(\rho - m - \tau)(\rho - m - \tau - 1) Q_{o,\tau} + (\rho - m - \tau) Q_{1,\tau} + Q_{2,\tau} \right]$$

pour $m \geqslant -\nu + 1$.

f. - Quand le point de l'infini est un point singulier irrégulier il n'existe pas de solutions régulières au voisinage de l'infini. Mais par contre il peut exister des solutions normales. Une solution de (A, 2) est dite normale si elle est de la forme :

(A, 11) $$R = V e^{\Omega}$$

V étant une solution régulière au sens stricte de l'équation transformée et Ω étant un polynome en r. Pour trouver les solutions normales à l'infini il faut donc déterminer les formes possibles de Ω telles qu'après le changement de fonction inconnue (A, 11) le point de l'infini soit au moins un point quasi-régulier. La théorie générale pour déterminer Ω existe mais nous n'aurons pas besoin de l'appliquer. En effet les équations du type (A, 1) auxquelles nous aurons besoin d'appliquer la transformation (A, 11) sont telles que la forme limite des coefficients M et N pour r tendant vers infini suggère immèdiatement la forme du polynome Ω.

APPENDICE B

Les états liés dans un potentiel newtonnien.

a. - Nous rappelons ici très rapidement les principales étapes de la résolution de l'équation de Schrödinger non relativiste pour un potentiel newtonnien. Il va sans dire que, du point de vue mathématique, la substitution du potentiel newtonnien au potentiel coulombien n'entraîne qu'une substitution des constantes qui figurent dans l'équation.

L'équation de Schrödinger pour un potentiel newtonnien est :

$$-\frac{\hbar^2}{2m}\Delta\Psi = \left(i\hbar\frac{\partial}{\partial t} - P\right)\Psi \tag{B, 1}$$

avec :

$$P = \frac{-kMm}{r}$$

Les solutions Ψ de l'équation (B, 1) qui sont en même temps fonctions propres des opérateurs : énergie et carré du moment cinétique :

$$i\hbar\frac{\partial\Psi}{\partial t} = W\Psi ;\quad \hbar^2\Lambda^2_{(\theta,\varphi)}\Psi = \hbar^2 l(l+1)\Psi ;\quad l \text{ entier} \geq 0$$

où

$$\Lambda^2_{(\theta,\varphi)} \equiv -\left[\frac{1}{\sin\theta}\frac{\partial}{\partial\theta}\left(\sin\theta\frac{\partial}{\partial\theta}\right) + \frac{1}{\sin^2\theta}\frac{\partial^2}{\partial\varphi^2}\right]$$

sont de la forme

$$\Psi = R(r)Y_l(\theta,\varphi)e^{-i\frac{W}{\hbar}t}$$

Y_l étant une harmonique sphérique et R étant une solution de l'équation radiale :

(B, 2) $$\frac{d^2R}{dr^2} + \frac{2}{r}\frac{dR}{dr} + \left[- \varkappa^2 + \frac{2kMm^2}{\hbar^2 r} - \frac{l(l+1)}{r^2} \right] R = 0$$

où :

(B, 3) $$\varkappa^2 = - \frac{2mW}{\hbar^2}$$

Pour les états liés, que nous nous bornons ici à considérer, $W < 0$ et par conséquent $\varkappa$ est rèel. Pour fixer les idées nous supposons $\varkappa > 0$.

b. - Par le changement d'inconnue

$$R = \frac{F}{r}$$

on ramène l'équation (B, 2) à la forme canonique :

$$\frac{d^2F}{dr^2} + \left[- \varkappa^2 + U(r, l) \right] F = 0$$

où $U(r, l)$, qu'on peut appeler le potentiel effectif, est

(B, 4) $$U(r, l) \equiv \frac{2kMm^2}{\hbar^2 r} - \frac{l(l+1)}{r^2}$$

Ce potentiel effectif qui ne dépend que de r et de l tend toujour à l'infini comme r^{-1}.

c. - Ecrivons l'équation (B, 2) sous la forme :

$$r^2 \frac{d^2R}{dr^2} + 2r \frac{dR}{dr} + \left[- \varkappa^2 r^2 + \frac{2kMm^2}{\hbar^2} r - l(l+1) \right] R = 0$$

On constate facilement que $r = 0$ est un point singulier régulier et que $r = \infty$ est un point singulier irrégulier.

Pour $r = \infty$ la forme limite de l'équation (B, 2) est

(B, 5) $$\frac{d^2R}{dr^2} - \varkappa^2 R = 0$$

qui admet des solutions de la forme

$$R = A e^{\varepsilon \varkappa r} \qquad \varepsilon^2 = +1 \qquad A : \text{const}$$

Ceci suggère donc de calculer des solutions normales à l'infini de la forme

(B, 6) $$R = V e^{-\varkappa r}$$

Seul $\varepsilon = -1$ a été rétenu pour assurer l'annulation de R pour $r = \infty$. Après le changement d'inconnue (B, 6), l'équation pour V est :

(B, 7) $$r^2 \frac{d^2V}{dr^2} + 2(1 - \varkappa r) r \frac{dV}{dr} + \left[2\left(\frac{kMm^2}{\hbar^2} - \varkappa\right) r - l(l+1) \right] V = 0$$

L'équation indicielle au point $r = \infty$ est maintenant :

(B, 8) $$\varkappa \rho - \left(\frac{kMm^2}{\hbar^2} - \varkappa\right) = 0$$

Le point $r = \infty$ est donc devenu un point singulier quasi-régulier et il peut exister au voisinage de l'infini une solution régulière de l'équation (B, 7) :

(B, 9) $$V = \sum_0^\infty b_r r^{\rho - n} \qquad b_0 \neq 0$$

ρ étant la solution de (B, 8).

d. - L'équation indicielle au point $r = 0$, qui est encore un point singulier régulier est :

$$s(s+1) - l(l+1) = 0$$

dont les deux solutions sont

$$s_1 = l \qquad s_2 = -(l+1)$$

ainsi :

$$s_1 - s_2 = 2l+1 > 0$$

et il existe certainement au voisinage de $r = 0$ une solution régulière au sens stricte de la forme :

(B, 10) $$V = \sum_{n=0}^{\rho-\ell} a_n r^{l+n}$$

Si la solution ayant le développement (B, 9) existe le domaine de convergence de la série (B, 9) sera tout le plan complexe de la variable r , sauf peut-être aux points $r = \infty$ et $r = 0$. De même le développement (B, 10) est convergent dans tout le plan complexe sauf peut-être pour $r = \infty$. Si nous exigeons à la function V solution de (B, 7) d'être régulière au sens stricte à la fois au voisinage de $r = \infty$ et $r = 0$, les deux développements (B, 9) et (B, 10) doivent coincider et par conséquent V doit être de la forme :

(B, 11) $$V = \sum_{n=0}^{\rho-\ell} a_n r^{l+n} \qquad a_o \neq 0 \quad a_\rho \neq 0$$

d'où il s'en suit en particulier que :

$$\rho = 1 + n'$$

avec $n' \geqslant 0$. Ceci est une condition nécessaire. En fait, on peut démontrer facilement que ρ entier $\geqslant 1$ est aussi une condition suffisante pour que la solution V de la forme (B, 11) existe. On sait que les polynomes ainsi obtenus sont les polynomes de Laguerre.

e. - Posons :

$$n = \rho + 1$$

De (B, 8) et (B, 3) il vient

(B, 12) $$W_n = - \lambda^2 \frac{mc^2}{2n^2} \qquad n = 1, 2, \ldots$$

où :

$$\lambda = \frac{kMm}{\hbar c}$$

Insistons sur le résultat que nous venons de rappeler.

Les solutions de l'équation radiale de l'équation de Schrödinger pour un potentiel newtonnien, correspondant aux différentes valeurs de l'énergie de la suite discrète d'états liés, sont des solutions normales à l'infini et régulières au sens stricte à l'origine appartenant dans les deux cas à des exposants entiers (resp. ρ et 1).

APPENDICE C

Etats liés relativistes pour un potentiel coulombien.

a. - En Relativité Restreinte l'intégrale d'action d'une particule de masse m et charge -e , dans le champ électrique d'une particule de masse M $\gg$ m et charge +e, est :

$$S = \int_{t_o}^{t_1} \left[-mc^2\sqrt{1-\frac{v^2}{c^2}} + \frac{e^2}{r} \right] dt$$

quand le repère est un repère galiléen d'origine M .

Les moments conjugués aux variables x^i sont :

$$p_i = \frac{mv^i}{\sqrt{1-\frac{v^2}{c^2}}}$$

et l'Hamiltonien H , ou E si nous nous réfèrons à une valeur particulière, est

$$H = E = \frac{mc^2}{\sqrt{1-\frac{v^2}{c^2}}} - \frac{e^2}{r} \tag{C, 1}$$

H = E est une constante du mouvement.

Des définitions de p_i et E il vient

$$(E + \frac{e^2}{r})^2 - c^2 \sum_1^3 p_i^2 = m^2 c^4$$

Si nous posons :

$$E = -\frac{\partial}{\partial t} \qquad p_i = \frac{\partial S}{\partial x^i}$$

l'équation précédente devient l'équation de Hamilton-Jacobi

$$(-\frac{\partial S}{\partial t} + \frac{e^2}{r})^2 - c^2 \sum_1^3 (\frac{\partial S}{\partial x^i})^2 = m^2 c^4$$

Et si nous appliquons le principe de correspondance :

$$E \longrightarrow i\hbar \frac{\partial}{\partial t} \qquad p_i \longrightarrow -i\hbar \frac{\partial}{\partial x^i}$$

nous obtenons l'équation de Schrödinger relativiste pour le système dynamique envisagé :

$$(-\hbar^2 c^2 \Delta + m^2 c^4)\psi = (-\hbar^2 \frac{\partial^2}{\partial t^2} + 2i\hbar \frac{e^2}{r} \frac{\partial}{\partial t} + \frac{e^4}{r^2})\psi$$

qui est l'équation d'onde relativiste qu'il convient de considérer quand la particule de masse m et change -e a un spin zéro.

La séparation des variables temporelle et angulaires se fait d'une manière identique à celle du problème de l'App. B. On arrive ainsi à des fonctions ψ de la forme :

$$\psi = R(r) Y_l(\theta, \varphi) e^{-i \frac{E}{\hbar} t}$$

R devant être una solution de l'équation radiale

(C, 2) $$\frac{d^2 R}{dr^2} + \frac{2}{r} \frac{dR}{dr} + \left[-\varkappa^2 + \frac{q}{r} + \frac{\gamma^2 - l(l+1)}{r^2} \right] R$$

où

(C, 3) $$\varkappa^2 = \frac{1}{c^2 \hbar^2} (m^2 c^4 - E^2) \quad q = \frac{2E}{c\hbar}\gamma \quad \gamma = \frac{e^2}{\hbar c}$$

b. - Par le changement d'inconnue :

$$R = \frac{F}{r}$$

on ramène l'équation(C, 2) à la forme canonique :

$$\frac{d^2F}{dr^2} + \left[-\mathcal{H}^2 + U(r, l, E) \right] F = 0$$

où cette fois le potentiel effectif est :

$$U(r, l, E) = \frac{q}{r} + \frac{\gamma^2 - l(l+1)}{r^2} \qquad \text{(C, 4)}$$

On remarquera deux particularités de ce potentiel effectif.

a) Il dépend de q et par conséquent de l'énergie E

b) Son comportement à l'infini est en r^{-1} sauf si $q = 0$, et par conséquent $E = 0$, auquel cas le comportement est en r^{-2}.

c. - Pour les états liés $\mathcal{H}$ est réel. Nous supposerons $\mathcal{H} > 0$.

Ecrivons l'équation (C, 2) sous la forme :

$$r^2 \frac{d^2R}{dr^2} + 2r \frac{dR}{dr} + \left[-\mathcal{H}^2 r^2 + qr + \gamma^2 - l(l+1) \right] R = 0 \qquad \text{(C, 5)}$$

On constate facilement que $r=0$ est un point singulier régulier et que $r=\infty$ est un point singulier irrégulier.

Pour $r=\infty$ la forme limite de l'équation (C, 2) est encore l'équation (B, 5). Nous sommes donc amenés pour les mêmes raisons que dans l'App. B à nous intéresser aux solutions normales de l'équation (C, 5) de la forme :

$$R(r) = V(r) e^{-\mathcal{H} r}$$

L'équation pour V est :

$$r^2 \frac{d^2V}{dr^2} + 2(1 - \mathcal{H} r) \frac{dV}{dr} + \left[(q - 2\mathcal{H}) r + \gamma^2 - l(l+1) \right] V = 0 \qquad \text{(C, 6)}$$

$r = \infty$ est pour cette équation un point singulier quasi-régulier auquel correspond l'équation indicielle :

(C, 7) $$2\mathcal{H}\rho + 2\mathcal{H} - q = 0$$

L'équation (C, 6) peut donc admettre une solution de la forme

$$V = \sum_{n=0}^{\infty} b_n r^{\rho - n}$$

ρ étant la solution de (C, 7).

d. - L'équation indicielle au point $r = 0$ qui est encore un point singulier régulier est

$$s^2 + s + \gamma^2 - l(l+1) = 0$$

qui admet les deux solutions :

(C, 8) $$s_A = -\frac{1}{2} + \frac{\varepsilon_A}{2} \cdot \sqrt{(2l+1)^2 - 4\gamma^2}; \quad A = 1, 2 \quad \varepsilon_1 = 1, \varepsilon_2 = -1$$

Nous avons :

$$s_1 - s_2 = \sqrt{(2l+1)^2 - 4\gamma^2}$$

qui si $4\gamma^2$ n'est pas un entier, comme c'est le cas si $-e$ est la charge de l'electron, ne peut certainement pas être un entier. L'équation (C, 6) admet donc au voisinage de l'origine deux solutions régulières au sens stricte de la forme

$$V = \sum_{n=0}^{\infty} a_n r^{s+n}$$

s étant s_1 on s_2.

Toutefois pour $l > 0$ on a $s_2 < -1$ et il faut exclure la solution correspondant à cette racine si l'on veut, par exemple que $|\psi^2| r^2$ soit fini à l'origine.

Pour $l=0$ par contre on a :

$$s_1(l=0) = -\frac{1}{2}+\frac{1}{2}\sqrt{1-4\gamma^2}$$

donc :

$$-1 < s_1(l=0) < 0$$

mais aussi [4] :

$$s_2(l=0) = -\frac{1}{2}-\frac{1}{2}\sqrt{1-4\gamma^2}$$

et par conséquent :

$$-1 < s_2(l=0) < 0$$

Aux deux racines $s_A(l=0)$ correspondent des solutions régulières aus sens stricte qui sont parfaitement acceptables.

Le même raisonnement de l'App. B permet de conclure que si on exige à la fonction V d'être régulière au voisinage de l'infini, régulière au voisinage de zéro, et telle que $r^2 V$ soit fini à l'origine, elle doit être nécessairement de la forme :

(C, 9) $$V = \sum_{n=0}^{\rho-s} a_n r^{s+n}$$

où $s = s_1$ si $l > 0$ et $s = s_1$ ou $s = s_2$ si $l = 0$. On tire en particulier de ce résultat que :

(C, 10) $$\rho = s + n' \qquad \text{avec } n' \text{ entier} \geq 0$$

Ceci est encore une condition nécessaire seulement. Mais on démontrerait aussi très facilement qu'étant donné ρ satisfaisant à la condition précédente la solution de la forme (C, 9) existe.

e. - Posons :

(C, 11) $$\nu = \rho + 1$$

Nous pouvons écrire, compte tenu de (C, 3), après élévation au carré l'équation (C, 7) sous la forme :

$$\nu^2(m^2c^4 - E^2) = E^2\gamma^2$$

d'où il vient :

$$(C,12) \qquad E_\nu = \frac{mc^2}{\sqrt{1+\frac{\gamma^2}{\nu^2}}}$$

Seule la détermination positive est solution de (C, 7). En effet γ est par définition positif et d'après (C, 10) et $s > -1$, $\rho+1$ est aussi positif, d'où q et par conséquent E positifs.

Il convient encore de compléter ce résultat par une discussion plus détaillée portant sur les valeurs possibles de V . De (C, 11), (C, 10), (C, 8) et la discussion qui a été faite sur les valeurs possibles de A , il vient :

$$\nu = n'+1 + \frac{1}{2}\left(\sqrt{(2l+1)^2 - 4\gamma^2} - 1\right)$$

pour $l > 0$, et pour $l = 0$ quand l'exposant à l'origine est s_1. Si nous développons en série la racine carrée (si -e est la charge de l'electron $\gamma \simeq \frac{1}{137}$) nous obtenons :

$$\nu = n' + 1 + l - \frac{\gamma^2}{2l+1}\left[1 + 0\left(\frac{\gamma}{2l+1}\right)^2\right]$$

D'autre part pour $l = 0$, quand l'exposant à l'origine est s_2 , nous avons :

$$\nu = n' + 1 - \frac{1}{2}\left(\sqrt{1 - 4\gamma^2} + 1\right)$$

et en développant la racine carrée :

$$\nu = n' + \gamma^2 + 0(\gamma^2)$$

Ainsi les valeurs possibles de ν sont :

$$\nu : \; 0+\delta_o, \; 1+\delta_1, \ldots, \; n+\delta_n \qquad n \text{ entier} \geq 0$$

où les δ_n sont des quantités qui sont au maximum de l'ordre de γ^2.

Supposons $n \geq 1$. Si nous négligeons les termes d'ordre supérieur à γ^2 dans (C, 12) nous obtenons

$$E_n = mc^2 - \gamma^2 \frac{mc^2}{2n^2} \tag{C, 13}$$

et nous rétrouvons au terme mc^2 près, comme il se devrait, les énergies non relativistes des états liés.

f. - Par contre si nous supposons $n = 0$ et négligeons encore les termes d'ordre supérieur à γ^2 nous obtenons :

$$E_o = mc^2\gamma$$

Cet état n'a pas d'analogue dans le cas non relativiste. On peut dire qu'il est un état super-lié.

Remarquons enfin que si nous avions supposé carrément $\delta_o = 0$, c'est-à-dire $\nu = 0$ et par conséquent $\rho = -1$, nous aurions obtenu

$$E_o = 0$$

qui comme nous l'avons vu, quand nous avons considéré les particularités du potentiel effectif (C, 4), était la seule valeur de l'énergie qui entraînait un comportement à l'infini en r^{-2} pour ce potentiel.

BIBLIOGRAPHIE

[1] Wittaker & Watson : A course of Modern Analysis. Fourth edition.

[2] A. R. Forsyth : Theory of differential equations. Dover publications.

[3] A. Perès : Phys. Rev. 120 (1960).

[4] R. M. Lauger e N. Rosen : Phys. Rev. 37 (1931)

[5] Singh & Pandey : Proc. Nat. Inst. Sci. India, A vol. 26 n. 6(1960).

CENTRO INTERNAZIONALE MATEMATICO ESTIVO

(C. I. M. E.)

G. FERRARESE

PROPRIETA' DI SECONDO ORDINE DI UN GENERICO RIFERIMENTO FISICO IN RELATIVITA' GENERALE.

PROPRIETA' DI SECONDO ORDINE DI UN GENERICO RIFERIMENTO FISICO IN RELATIVITA' GENERALE.

In questo seminario vengono estese ad un generico riferimento fisico della relatività generale alcune proprietà di secondo ordine recentemente stabilite per una congruenza rigida[(1)]. Per brevità, ciò che non menoma la generalità, si fa uso della tecnica delle proiezioni in coordinate adattate al riferimento fisico[(2)]; tecnica che viene rielaborata nella prima parte alla luce dei metodi dei riferimenti anolonomi[(3)].

L'estensione si consegue utilizzando la decomposizione naturale del tensore di curvatura[(4)], decomposizione che qui viene ritrovata direttamente a partire da una espressione intrinseca generale del tensore di Riemann.

Particolarmente significativa è una relazione in termini finiti tra il tensore vortice spaziale del riferimento, il tensore di deformazione e i due tensori di curvatura spazio-temporale e spaziale rispettivamente [cfr. (41)] la quale sembra particolarmente adatta a riconoscere la possibilità di eventuali moti rigidi in una V_4 curva. (Negli spazi piatti vale, come è noto, il teorema di Herglotz-Noether secondo cui le congruenze rigide sono tutte e sole quelle definenti un gruppo di isometrie ad un parametro. Tale teorema viene qui ritrovato come conseguenza quasi immediata della relazione sopra citata).

---

(1) Cfr. Pirani, F. A. E. e Williams, G., Rigid motion in a gravitational field, Séminaire Janet, 5e année n. 8-9 (1961-62).

(2) Cfr. Cattaneo, C., Proiezioni naturali e derivazione trasversa in una varietà riemanniana a metrica iperbolica normale, Annali di Matem. (IV), V. 48, p. 361(1959).

(3) Cfr. ad es. Schouten, J. A., Ricci calculus, 2ª ed. Springer-Verlag, Berlin-Gottingen-Heidelberg (1954).

(4) Cfr. Cattaneo Gasparini, I., Projections naturelles des tenseurs de courbure d'une varieté V_{n+1} à métrique hyperbolique normale, C. R. Acad. Sc. Paris, t. 252, p. 3722(1961).

La medesima relazione vale anche a ritrovare rapidamente alcune notevoli proprietà dei moti rigidi in uno spazio di Minkowski.

1. Coordinamento di alcune nozioni sui riferimenti anolonomi. Siano: V_4 una varietà differenziabile di classe sufficientemente elevata, riferita a coordinate locali (x^i) $(i = 0,1,2,3)$; $\{e_i\}$ e $\{\underset{\vee}{\theta}^i\}$ rispettivamente la base e la cobase naturali :

$$(1)\qquad e_i = \frac{\partial OP}{\partial x^i}, \quad \underset{\vee}{\theta}^i(e_h) = \delta^i_h \qquad (i, h = 0,1,2,3);$$

$\underset{\vee}{\overset{r}{\lambda}}$ quattro forme lineari indipendenti assegnate :

$$(2)\qquad \underset{\vee}{\overset{r}{\lambda}} = \overset{r}{\lambda}_i \underset{\vee}{\theta}^i \qquad (r = 0,1,2,3 \text{ indice ordinale});$$

$\underset{r}{\lambda}^i$ il reciproco di $\overset{r}{\lambda}_i$ nella matrice $\|\overset{r}{\lambda}_i\|$:

$$(3)\qquad \overset{r}{\lambda}_i \underset{s}{\lambda}^i = \delta^r_s, \quad \overset{r}{\lambda}_h \underset{r}{\lambda}^i = \delta^i_h .$$

Le quantità numeriche $\underset{r}{\lambda}^i$, ove l'indice in basso è ordinale e l'indice in alto i è tensoriale[5], individuano la base duale di $\{\underset{\vee}{\overset{r}{\lambda}}\}$, generalmente anolonoma, costituita dai quattro vettori contravarianti

$$(2')\qquad \underset{r}{\lambda} = \underset{r}{\lambda}^i e_i \qquad (r = 0,1,2,3)$$

(tetrade generica).

Per un generico vettore v (o più generalmente per un tensore) dello spazio tangente T_x si chiamano intrinseche le componenti secondo la base $\{\underset{r}{\lambda}\}$. Esse vengono contrassegnate con un indice in alto ($\overset{r}{v}$), per distinguerle dalle componenti naturali (v^i) che figurano nella decomposizione secondo la base $\{e_i\}$. Il passaggio dalle une alle altre è immediato, avendosi $\overset{r}{v} = \overset{r}{\lambda}_i v^i$ e inversamente $v^i = \underset{r}{\lambda}^i \overset{r}{v}$.

Oltre alla base (2'), le quantità $\underset{r}{\lambda}^i$ definiscono i quattro operatori dif-

---

(5) Si conviene di porre in basso o in alto gli indici ordinali e a lato gli indici tensoriali.

ferenziali (lineari omogenei) indipendenti

$$\partial_r = \lambda^i_r \frac{\partial}{\partial x^i} \qquad (r, i = 0, 1, 2, 3), \tag{4}$$

le derivate pfaffiane secondo i vettori covarianti (2').

Introdotte le corrispondenti parentesi di Poisson $[\partial_r, \partial_s] \equiv \partial_r\partial_s - \partial_s\partial_r$, si può scrivere

$$\partial_r\partial_s - \partial_s\partial_r = A^k_{rs}\,\partial_k \tag{5}$$

ponendo semplicemente

$$A^k_{rs} = (\partial_r \lambda^i_s - \partial_s \lambda^i_r)\lambda^k_i \equiv -\lambda^h_r \lambda^i_s \left(\frac{\partial \lambda^k_i}{\partial x^h} - \frac{\partial \lambda^k_h}{\partial x^i}\right). \tag{6}$$

Dall'ultima espressione del sistema triplo considerato appare evidente che : a) le singole quantità A^k_{rs} non dipendono dal sistema di coordinate adottato, ma soltanto dalle forme considerate (2); b) le quantità A^k_{rs}, antisimmetriche rispetto agli indici r ed s, si annullano se e solo se esistono quattro funzioni $y^r(x)$ (r = 0, 1, 2, 3) tali che sia $\lambda^r_i = \frac{\partial y^r}{\partial x^i}$, con che $\partial_r \equiv \frac{\partial}{\partial y^r}$; si annullano cioè se e solo se le forme differenziali $\lambda^r_i dx^i$ (r = 0, 1, 2, 3) sono integrabili.

Le quantità A^k_{rs} definiscono pertanto in modo intrinseco un tensore triplo, il cosidetto tensore di anolonomia della base $\{\lambda^r\}$:

$$\mathbf{A} = A^k_{rs}\, \lambda^r \otimes \lambda^s \otimes \lambda_k . \tag{6'}$$

Tale tensore caratterizza, col suo annullarsi, le basi naturali.

Le quantità A^k_{rs} sono naturalmente subordinate, oltre che alle proprietà di antisimmetria $A^k_{rs} = -A^k_{sr}$, alle identità di Jacobi

$$[[\partial_r, \partial_s], \partial_n] + [[\partial_s, \partial_n], \partial_r] + [[\partial_n, \partial_r], \partial_s] = 0 \tag{7}$$

che, avuto riguardo alla (5), si scrivono

$$\partial_n A^k_{rs} + \partial_r A^k_{sn} + \partial_s A^k_{nr} = A^m_{rs} A^k_{mn} + A^m_{sn} A^k_{mr} + A^m_{nr} A^k_{ms} . \tag{7'}$$

Se la varietà V_H è dotata di metrica $ds^2 \equiv g_{ih}\, dx^i dx^h$ (quindi di connessione riemanniana) e si adotta un riferimento anolonomo $\{\lambda_r\}$, in luogo dell'abituale riferimento naturale, intervengono naturalmente[(6)] le componenti intrinseche del tensore metrico

(8) $$g_{rs} = \lambda^i_r \lambda^h_s g_{ih} \qquad \text{ovvero} \qquad g^{rs} = \lambda^r_i \lambda^s_h g^{ih} ,$$

nonchè il sistema multiplo (di cui si richiama appresso il significato)

(9) $$\mathcal{R}_{rsn} = [rs,n] + \frac{1}{2}\left(A_{nrs} + A_{rsn} - A_{snr} \right)$$

ove

(10) $$A_{nrs} = A^{k}_{nr}\, g_{ks} \quad , \quad [rs,n] = \frac{1}{2}\left(\partial_r g_{sn} + \partial_s g_{nr} - \partial_n g_{rs} \right) .$$

Le quantità $[rs,n]$ sono del tutto analoghe ai simboli di Christoffel di 1ª specie, salvo la sostituzione delle g_{ih} con le g_{rs} e della derivazione ordinaria $\partial/\partial x^i$ con la derivazione pfaffiana ∂_r, e la parte antisimmetrica (rispetto ai primi due indici) di $\mathcal{R}_{rsn}$ è A_{rsn} :

(11) $$\mathcal{R}_{rsn} - \mathcal{R}_{srn} = A_{rsn} .$$

Ciò posto, la derivazione intrinseca riemanniana di un tensore T si ottiene, come è noto, sostituendo, nell'espressione della derivata covariante ordinaria, alle componenti naturali del tensore le componenti intrinseche, alla derivazione parziale ordinaria la derivazione pfaffiana e ai simboli di Christoffel i coefficienti

(9') $$\mathcal{R}^{n}_{rs} = g^{mn} \mathcal{R}_{rsm} .$$

Per quanto riguarda il tensore di curvatura basta osservare che anche in forma intrinseca, l'identità di Ricci si scrive al modo usuale :

(6) Cfr. ad es. G. Ferrarese, Sulle equazioni di moto di un sistema soggetto a un vincolo anolonomo mobile, Rend. di Matem. , Vol. XXII, 3-4 (1963), n. 1, 2, 3, 4.

$$\nabla_s \nabla_r v_n = \nabla_r \nabla_s v_n - v_m \overset{m}{R}_{nrs} \tag{12}$$

ove ∇_r indica la derivazione covariante intrinseca :

$$\nabla_r v_n = \partial_r v_n - \overset{k}{\mathcal{R}}_{rn} v_k \; . \tag{13}$$

Esplicitando il 1° membro della (12) si ritrova facilmente, per il tensore di Riemann, l'espressione generale[7]

$$\overset{m}{R}_{nrs} = \partial_s \overset{m}{\mathcal{R}}_{rn} + \overset{k}{\mathcal{R}}_{rn} \overset{m}{\mathcal{R}}_{sk} - \partial_r \overset{m}{\mathcal{R}}_{sn} - \overset{k}{\mathcal{R}}_{sn} \overset{m}{\mathcal{R}}_{rk} + \overset{k}{A}_{rs} \overset{m}{\mathcal{R}}_{kn} \tag{14}$$

ove l'ultimo termine, non classico, è diretta conseguenza della anolonomia del riferimento[8].

2. Riferimenti anolonomi naturalmente associati ad un campo di faccette[9].

Particolarmente interessante per la relatività generale è il caso in cui in V_4, dotata della sola struttura di varietà differenziabile, sia fissato un campo di faccette Σ_x , cioè una sola forma lineare γ_i , definita a meno di un fattore di proporzionalità. (Se la varietà V_4 è dotata di metrica iperbolica normale, il campo di faccette Σ_x , purchè del genere spazio, definisce un riferimento fisico, per altro rappresentato, in tal caso,anche dalla congruenza dei vettori normali alle Σ_x).

Supposto $\gamma_0 \neq 0$, conviene fissare l'attenzione sul riferimento

$$\overset{0}{\lambda} \equiv -\gamma_i \theta^i \; , \quad \overset{\alpha}{\lambda} \equiv \theta^\alpha \qquad (\alpha = 1, 2, 3)^{(10)}, \tag{15}$$

---

(7) Cfr. loc. cit. (2), p. 172 e, per connessioni lineari generiche, A. Lichnerowicz, Theorie globale des connexions et des groupes d'holonomie, Roma, Cremonese (1955), p. 87.

(8) Si noti che il tensore di anolonomia compare nella (14) anche per il tramite dei coefficienti $\overset{n}{\mathcal{R}}_{rs}$ [cfr. (9') e (9)] .

(9) Per il caso di una varietà riemanniana si confronti (1).

(10) Si conviene che gli indici greci varino da 1 a 3, quelli latini da 0 a 3.

(generalmente anolonomo e dipendente, oltre che dal campo di faccette Σ_x, dal sistema di coordinate prescelto) nonchè sulla base duale che risulta definita dai vettori contravarianti

$$\underset{0}{\lambda} = -\frac{1}{\gamma_0} e_0 \quad , \quad \underset{\alpha}{\lambda} = e_\alpha - \frac{\gamma_\alpha}{\gamma_0} e_0 \, . \tag{15'}$$

Viene così realizzata in ogni punto di V_4, tanto per lo spazio tangente, quanto per il suo duale, una decomposizione nel prodotto di uno spazio unidimensionale per uno spazio tridimensionale. Precisamente $\underset{0}{\lambda}$ costituisce una base dello spazio Θ_x dei vettori tangenti alle linee x^0 = var. e la terna $\underset{\alpha}{\lambda}$ (α = 1, 2, 3) una base per i vettori appartenenti alla faccetta Σ_x, cioè per i vettori s^i che soddisfano alla condizione $s^i \gamma_i = 0$. La (15') definisce pertanto una decomposizione dello spazio tangente T_x nella somma di Θ_x e Σ_x: $T_x = \Theta_x + \Sigma_x$; la (15) una decomposizione analoga per lo spazio duale di T_x, e così via per gli spazi tensoriali associati.

Ad esempio per un vettore $\boldsymbol{v}$, ove si consideri la rappresentazione intrinseca secondo la base (15') : $\boldsymbol{v} = \overset{r}{v} \underset{r}{\lambda}$, si ha la decomposizione, $\boldsymbol{v} = A + B$ nelle due parti

$$\begin{aligned} A &= \overset{0}{v} \underset{0}{\lambda} \quad , \quad \overset{0}{v} = \overset{0}{\lambda}_i v^i = -\gamma_i v^i \\ B &= \overset{\alpha}{v} \underset{\alpha}{\lambda} \quad , \quad \overset{\alpha}{v} = \overset{\alpha}{\lambda}_i v^i = v^\alpha \qquad (\alpha = 1, 2, 3) \, . \end{aligned} \tag{16}$$

Per un tensore doppio contravariante $\boldsymbol{T} = \overset{rs}{T} \underset{r}{\lambda} \otimes \underset{s}{\lambda}$ si ha la decomposizione $\boldsymbol{T} = \boldsymbol{T}_1 + \boldsymbol{T}_2 + \boldsymbol{T}_3 + \boldsymbol{T}_4$ nelle quattro parti

$$
(17)\qquad \begin{cases}
T_1 = \overset{00}{T}\, \underset{0}{\lambda} \otimes \underset{0}{\lambda}\ , & \overset{00}{T} = \overset{0}{\lambda}_i \overset{0}{\lambda}_h T^{ih} = \gamma_i \gamma_h T^{ih} \\
T_2 = \overset{0\alpha}{T}\, \underset{0}{\lambda} \otimes \underset{\alpha}{\lambda}\ , & \overset{0\alpha}{T} = \overset{0}{\lambda}_i \overset{\alpha}{\lambda}_h T^{ih} = -\gamma_i T^{i\alpha} \\
T_3 = \overset{\alpha 0}{T}\, \underset{\alpha}{\lambda} \otimes \underset{0}{\lambda}\ , & \overset{\alpha 0}{T} = \overset{\alpha}{\lambda}_i \overset{0}{\lambda}_h T^{ih} = -\gamma_h T^{\alpha h} \\
T_4 = \overset{\alpha\beta}{T}\, \underset{\alpha}{\lambda} \otimes \underset{\beta}{\lambda}\ , & \overset{\alpha\beta}{T} = \overset{\alpha}{\lambda}_i \overset{\beta}{\lambda}_h T^{ih} = T^{\alpha\beta}
\end{cases}
$$

(decomposizione naturale) ecc. ecc.

Passando a calcolare le componenti non nulle del tensore di anolonomia della base (15) si trova, in base alla (6) :

$$
(18)\qquad \underset{\alpha\beta}{\overset{0}{A}} = \gamma_0 \left(\tilde{\partial}_\alpha \frac{\gamma_\beta}{\gamma_0} - \tilde{\partial}_\beta \frac{\gamma_\alpha}{\gamma_0} \right) , \qquad \underset{0\alpha}{\overset{0}{A}} = \tilde{\partial}_\alpha \log \gamma_0 + \gamma_0 \bar{\partial} \frac{\gamma_\alpha}{\gamma_0}
$$

ove

$$
(19)\qquad \bar{\partial} \equiv \partial_0 = -\frac{1}{\gamma_0} \frac{\partial}{\partial x^0} , \qquad \tilde{\partial}_\alpha \equiv \partial_\alpha = \frac{\partial}{\partial x^\alpha} - \frac{\gamma_\alpha}{\gamma_0} \frac{\partial}{\partial x^0} .
$$

(derivata longitudinale e trasversa rispettivamente).

Nel seguito si pone (conservando, sia pure con significato diverso, le notazioni abituali per il tensore vortice spaziale e per il vettore di curvatura del caso riemanniano) :

$$
(20)\qquad \tilde{\Omega}_{\alpha\beta} = \underset{\alpha\beta}{\overset{0}{A}} , \qquad C_\alpha = \underset{0\alpha}{\overset{0}{A}} .
$$

Si noti esplicitamente che una trasformazione di coordinate del tipo

$$
(21)\qquad x^{0'} = \psi(x^0, x^1, x^2, x^3) , \qquad x^{\alpha'} = x^\alpha
$$

(interna alla congruenza x^0 = var.) lascia invariata la base anolonoma (15), ovvero (15'), sì che essa conserva (salvo l'aggiunta di un apice) le espressioni (18) delle singole quantità $\tilde{\Omega}_{\alpha\beta}$ e C_α . Disponendo della funzione $\psi(x)$ non sarebbe pertanto restrittivo (fissate che siano il campo di faccette Σ_x e la congruenza x^0 = var.) supporre γ_0, ad esempio, funzione assegnata delle x^β, purchè non nulla.

I due tensori $\underset{\vee}{\Omega} \equiv \tilde{\Omega}_{\alpha\beta} \overset{\alpha}{\underset{\vee}{\tilde{\partial}}} \otimes \overset{\beta}{\underset{\vee}{\tilde{\partial}}}$ e $\underset{\vee}{C} \equiv C_\alpha \overset{\alpha}{\underset{\vee}{\tilde{\partial}}}$, appartenenti a Σ_x^* (duale di Σ_x) godono, come nel caso di una varietà riemanniana, di espressive proprietà geometriche. L'annullarsi di entrambi, quindi del tensore di anolonomia, è condizione necessaria e sufficiente perchè il riferimento (15) sia olonomo, ovvero integrabile la forma differenziale $\gamma_i dx^i$. L'annullarsi del solo tensore $\underset{\vee}{\Omega}$ (che dipende dai rapporti γ_α/γ_0) è caratteristico dei campi di faccette Σ_x raccordabili su ipersuperficie.

Per quanto riguarda $\underset{\vee}{C}$ (il quale dipende, oltre che dal campo di faccette Σ_x, dalla congruenza x^0 = var.) la condizione $\underset{\vee}{C}$ = grad $\varphi \equiv \frac{\partial \varphi}{\partial x^\alpha} \overset{\alpha}{\underset{\vee}{\tilde{\partial}}}$ con $\varphi = \varphi(x^1, x^2, x^3)$ è necessaria e sufficiente perchè il campo di faccette Σ_x sia stazionario:

$$\frac{\partial \gamma_i}{\partial x^0} = 0$$

Se si ha insieme $\underset{\vee}{\Omega} = 0$, $\underset{\vee}{C} = \widetilde{\text{grad}}\, \varphi \equiv \tilde{\partial}_\alpha \varphi \overset{\alpha}{\underset{\vee}{\tilde{\partial}}}$ la forma differenziale $\gamma_i dx^i$ ammette un fattore integrante.

Per quanto riguarda la (5) essa dà luogo, in base alla (20), alle due formule di commutazione

$$\tilde{\partial}_\alpha \tilde{\partial}_\beta - \tilde{\partial}_\beta \tilde{\partial}_\alpha = \tilde{\Omega}_{\alpha\beta} \bar{\partial} \quad , \quad \bar{\partial} \tilde{\partial}_\alpha - \tilde{\partial}_\alpha \bar{\partial} = C_\alpha \bar{\partial} \tag{22}$$

mentre l'identità di Jacobi (7') si traduce in

$$\begin{cases} \tilde{\partial}_\nu \tilde{\Omega}_{\rho\sigma} + \tilde{\partial}_\rho \tilde{\Omega}_{\sigma\nu} + \tilde{\partial}_\sigma \tilde{\Omega}_{\nu\rho} = C_\nu \tilde{\Omega}_{\rho\sigma} + C_\rho \tilde{\Omega}_{\sigma\nu} + C_\sigma \tilde{\Omega}_{\nu\rho} \\ \tilde{\partial}_\rho C_\sigma - \tilde{\partial}_\sigma C_\rho = \bar{\partial} \tilde{\Omega}_{\rho\sigma} \end{cases} \tag{23}$$

3. Caso riemanniano.

Qui e nel seguito si suppone che la varietà V_4, finora semplicemente differenziabile, sia dotata di una metrica, $ds^2 \equiv g_{ih}\, dx^i dx^h$ di tipo iperbolico normale (-+++) e che le linee x^0 = var. siano del genere tempo ($g_{00} < 0$). La metrica può essere introdotta assegnando le componenti intrinseche

(24) $$g_{rs} = \underset{r}{\lambda} \cdot \underset{s}{\lambda} \equiv \underset{r}{\lambda}{}^i \underset{s}{\lambda}{}^h g_{ih}$$

secondo il riferimento anolonomo (15') naturalmente associato al campo di faccette Σ_x.

Si suppone altresì che le linee x^0 = var., tangenti a $\underset{0}{\lambda}$, siano ortogonali in ogni punto a Σ_x, quindi $\underset{0}{\lambda} \cdot \underset{\alpha}{\lambda} = 0$ e che insieme risulti (disponendo del fattore di proporzionalità per γ_i, inessenziale per i vettori $\underset{\alpha}{\lambda}$ ma disponibile per $\underset{0}{\lambda}$) $\underset{0}{\lambda} \cdot \underset{0}{\lambda} = -1$. Per fissare completamente la metrica non resta che assegnare i prodotti $\underset{\alpha}{\lambda} \cdot \underset{\beta}{\lambda} = \gamma_{\alpha\beta}(x)$ (α, β = 1, 2, 3), naturalmente con la condizione che la forma ternaria $\gamma_{\alpha\beta}\, x^\alpha x^\beta$ sia definita positiva.

Si avrà allora

(24') $$g_{00} = -1\,, \quad g_{0\alpha} = 0\,, \quad g_{\alpha\beta} = \gamma_{\alpha\beta} \qquad (\alpha, \beta = 1, 2, 3)$$

nonchè, per gli elementi reciproci,

(24'') $$g^{00} = -1\,, \quad g^{0\alpha} = 0\,, \quad g^{\alpha\beta} = \gamma^{\alpha\beta} \qquad (\alpha, \beta = 1, 2, 3)$$

ove $\gamma^{\alpha\beta}$ è il reciproco di $\gamma_{\alpha\beta}$ nella matrice di terzo ordine $\| \gamma_{\alpha\beta} \|$: $\gamma_{\alpha\rho}\gamma^{\beta\rho} = \delta_\alpha^\beta$.

Dalle formule $g_{ih} = \overset{r}{\lambda}_i \overset{s}{\lambda}_h g_{rs}$ in cui, come dalla (15), $\overset{0}{\lambda}_i = -\gamma_i$,

$\overset{\alpha}{\lambda}_i = \delta_i^\alpha$ segue poi, per le componenti naturali del tensore metrico,

(25) $$g_{0i} = -\gamma_0 \gamma_i \quad , \quad g_{\alpha\beta} = \gamma_{\alpha\beta} - \gamma_\alpha \gamma_\beta$$

nonchè

(25') $$g^{00} = \frac{1}{\gamma_0^2}(\gamma_\alpha \gamma_\beta \gamma^{\alpha\beta} - 1) , \quad g^{0\alpha} = -\frac{\gamma_\beta}{\gamma_0}\gamma^{\alpha\beta} , \quad g^{\alpha\beta} = \gamma^{\alpha\beta} .$$

Specificata così la metrica, dalla (9') si trae, avuto riguardo alle (9) e (18), che i soli coefficienti non nulli della connessione riemanniana secondo la base (15) sono :

(26) $$\begin{cases} \overset{0}{\underset{0\alpha}{R}} = C_\alpha , \quad \overset{0}{\underset{\alpha\beta}{R}} = \mathcal{E}_{\alpha\beta} , \quad \overset{\delta}{\underset{\alpha\beta}{R}} = \left\{ {\delta \atop \alpha\beta} \right\} \\ \overset{\alpha}{\underset{00}{R}} = C^\alpha \equiv \gamma^{\alpha\beta} C_\beta , \quad \overset{\beta}{\underset{0\alpha}{R}} = \overset{\beta}{\underset{\alpha 0}{R}} = \mathcal{E}_\alpha^\beta \equiv \gamma^{\beta\rho} \mathcal{E}_{\alpha\rho} \end{cases}$$

ove si è posto

(27) $$\mathcal{E}_{\alpha\beta} = \frac{1}{2}(\tilde{K}_{\alpha\beta} + \tilde{\Omega}_{\alpha\beta}) , \quad \tilde{K}_{\alpha\beta} = \bar{\partial}\gamma_{\alpha\beta} , \quad \left\{ {\delta \atop \alpha\beta} \right\} = \frac{1}{2}\gamma^{\delta\rho}(\tilde{\partial}_\alpha \gamma_{\beta\rho} + \tilde{\partial}_\beta \gamma_{\rho\alpha} - \tilde{\partial}_\rho \gamma_{\alpha\beta}) .$$

Come il tensore di anolonomia del riferimento (15) si riassume nei due tensori $\tilde{\Omega}_{\alpha\beta}$ (vortice spaziale) e C_α (vettore di curvatura delle linee x^0 = var.), così la parte $[rs, n]$ della connessione (9) si esaurisce nei simboli di Christoffel spaziali $\widetilde{(\alpha\beta, \rho)} = \frac{1}{2}(\tilde{\partial}_\alpha \gamma_{\beta\rho} + \tilde{\partial}_\beta \gamma_{\rho\alpha} - \tilde{\partial}_\rho \gamma_{\alpha\beta})$ e nel tensore di deformazione $\tilde{K}_{\alpha\beta} = \bar{\partial}\gamma_{\alpha\beta}$. Si noti anzi esplicitamente che nella (26) i due tensori spaziali $\tilde{K}_{\alpha\beta}$ e $\tilde{\Omega}_{\alpha\beta}$ non vi compaiono singolarmente ma solo attraverso la combinazione $\mathcal{E}_{\alpha\beta} = \frac{1}{2}(\tilde{K}_{\alpha\beta} + \tilde{\Omega}_{\alpha\beta})$.

Con la determinazione dei coefficienti della connessione risulta immediata la decomposizione naturale del derivato covariante di un vettore o tensore qualunque. Ad esempio, per un vettore covariante di componenti intrinseche

v_s ($v_0 = -\frac{1}{\gamma_0} v_0$, $v_\alpha = v_\alpha - \frac{\gamma_\alpha}{\gamma_0} v_0$) da $\nabla_r v_s = \partial_r v_s - R_{rs}^{\ \ \ k} v_k$ si traggono automaticamente le quattro proiezioni naturali

$$
(28) \qquad \begin{cases} \nabla_\rho v_\sigma = \tilde{\nabla}_\rho v_\sigma - \xi_{\rho\sigma} v_0 \, , & \nabla_\rho v_0 = \tilde{\partial}_\rho v_0 - \xi_\rho^{\ \sigma} v_\sigma \\ \nabla_0 v_\sigma = \bar{\partial}_0 v_\sigma - \xi_\sigma^{\ \rho} v_\rho - C_\sigma v_0 \, , & \nabla_0 v_0 = \bar{\partial}_0 v_0 - C^\rho v_\rho \end{cases}
$$

ove $\tilde{\nabla}$ è il simbolo di derivazione covariante trasversa[11]:

$$
(29) \qquad \tilde{\nabla}_\rho v_\sigma = \tilde{\partial}_\rho v_\sigma - \left\{ {\delta \atop \rho\sigma} \right\} v_\delta \, .
$$

Per un vettore spaziale, cioè appartenente a Σ_x ($v_0 = 0$, $v_\alpha = v_\alpha$), sussiste l'identità $\nabla_\rho v_\sigma \equiv \tilde{\nabla}_\rho v_\sigma$ e, per le rimanenti componenti del tensore $\nabla_r v_s$, generalmente non nulle, si ha[12]

$$
(30) \qquad \nabla_\rho v_0 = -\xi_\rho^{\ \sigma} v_\sigma \, , \quad \nabla_0 v_\sigma = \bar{\partial}_0 v_\sigma - \xi_\sigma^{\ \rho} v_\rho \, , \quad \nabla_0 v_0 = -C^\rho v_\rho \, .
$$

Naturalmente il teorema di Ricci $\nabla_r g_{sn} = 0$ dà luogo ad un teorema analogo per il tensore metrico spaziale $\gamma_{\alpha\beta}$, come risulta dalla (24') :

$$
(31) \qquad \nabla_\rho g_{\sigma\nu} \equiv \tilde{\nabla}_\rho \gamma_{\sigma\nu} = 0 \, .
$$

4. Decomposizione naturale del tensore di curvatura. Identità del Bianchi.

Per quanto riguarda il tensore di curvatura, dalla espressione generale (14), avuto riguardo alle (20) e (26), seguono le componenti significative

---

(11) Cfr. loc. cit. (1), p. 371.

(12) Cfr. loc. cit. (1), p. 384.

$$
(32) \qquad \begin{cases} \overset{\mu}{R}_{\nu\rho\sigma} = \overset{\mu}{P}_{\nu\rho\sigma} + \xi_{\rho\nu}\,\xi_{\sigma}{}^{\mu} - \xi_{\sigma\nu}\,\xi_{\rho}{}^{\mu} + \hat{\Omega}_{\rho\sigma}\,\xi_{\nu}{}^{\mu} \\ \overset{0}{R}_{\nu\rho\sigma} = \tilde{\nabla}_{\sigma}\,\xi_{\rho\nu} - \tilde{\nabla}_{\rho}\,\xi_{\sigma\nu} + \tilde{\Omega}_{\rho\sigma}\,C_{\nu} \\ \overset{0}{R}_{\nu 0\sigma} = \tilde{\nabla}_{\sigma} C_{\nu} + C_{\sigma} C_{\nu} - \bar{\partial}\,\xi_{\sigma\nu} + \xi_{\nu}{}^{\rho}\,\xi_{\sigma\rho} \end{cases}
$$

ove le quantità

$$
(33) \qquad \overset{\mu}{P}_{\nu\rho\sigma} \equiv \tilde{\partial}_{\sigma}\widetilde{\left\{ {\mu \atop \rho\nu} \right\}} + \widetilde{\left\{ {\delta \atop \rho\nu} \right\}}\widetilde{\left\{ {\mu \atop \sigma\delta} \right\}} - \tilde{\partial}_{\rho}\widetilde{\left\{ {\mu \atop \sigma\nu} \right\}} - \widetilde{\left\{ {\delta \atop \sigma\nu} \right\}}\widetilde{\left\{ {\mu \atop \rho\delta} \right\}}
$$

definiscono un tensore spaziale del tutto analogo ad un tensore di Riemann costruito con l'impiego della metrica spaziale $\gamma_{\alpha\beta}$ e della derivazione trasversa.

In forma totalmente covariante, concordemente alle proiezioni calcolate per via diversa da I. Cattaneo Gasparini[13], si ha

$$
(32') \qquad \begin{cases} R_{\mu\nu\rho\sigma} = P_{\mu\nu\rho\sigma} + \xi_{\rho\nu}\,\xi_{\sigma\mu} - \xi_{\sigma\nu}\,\xi_{\rho\mu} + \tilde{\Omega}_{\rho\sigma}\,\xi_{\nu\mu} \\ R_{0\nu\rho\sigma} = -\tilde{\nabla}_{\sigma}\,\xi_{\rho\nu} + \tilde{\nabla}_{\rho}\,\xi_{\sigma\nu} - \tilde{\Omega}_{\rho\sigma}\,C_{\nu} \\ R_{0\nu 0\sigma} = \bar{\partial}\,\xi_{\sigma\nu} - \xi_{\nu}{}^{\rho}\,\xi_{\sigma\rho} - \tilde{\nabla}_{\sigma} C_{\nu} - C_{\sigma} C_{\nu} \end{cases}
$$

essendo naturalmente

$$
(33') \qquad P_{\mu\nu\rho\sigma} \equiv \gamma_{\mu\delta}\overset{\delta}{P}_{\nu\rho\sigma} = \tilde{\partial}_{\sigma}(\widetilde{\rho\nu,\mu}) - \tilde{\partial}_{\rho}(\widetilde{\sigma\nu,\mu}) + \gamma^{\varepsilon\delta}\left[(\widetilde{\sigma\nu,\varepsilon})(\widetilde{\rho\mu,\delta}) - (\widetilde{\rho\nu,\varepsilon})(\widetilde{\sigma\mu,\delta})\right].
$$

Si noti tuttavia che il tensore $P_{\mu\nu\rho\sigma}$ non gode generalmente di tutte le proprietà algebriche di un tensore di curvatura. Si ha infatti dalla (33')

(13) Cfr. loc. cit. (3), rilevando la diversa convenzione di segno qui assunta per il tensore di Riemann.

$$(34)\quad \begin{cases} P_{\mu\nu\rho\sigma} + P_{\mu\nu\sigma\rho} = 0 \quad , \quad P_{\mu\nu\rho\sigma} + P_{\nu\mu\rho\sigma} = -\tilde{K}_{\mu\nu}\tilde{\Omega}_{\rho\sigma} \\ P_{\mu\nu\rho\sigma} - P_{\rho\sigma\mu\nu} = \frac{1}{3}(\tilde{\Omega}_{\mu\nu}\tilde{K}_{\rho\sigma} - \tilde{K}_{\mu\nu}\tilde{\Omega}_{\rho\sigma} + \tilde{K}_{\mu\rho}\tilde{\Omega}_{\sigma\nu} + \tilde{K}_{\mu\sigma}\tilde{\Omega}_{\nu\rho} \\ \qquad\qquad - \tilde{\Omega}_{\mu\rho}\tilde{K}_{\sigma\nu} + \tilde{\Omega}_{\mu\sigma}\tilde{K}_{\nu\rho}) \\ P_{\mu\nu\rho\sigma} + P_{\mu\rho\sigma\nu} + P_{\mu\sigma\nu\rho} = 0 \ ; \end{cases}$$

ciò che suggerisce di assumere come tensore di curvatura spaziale il seguente :

$$(35)\quad \tilde{R}_{\mu\nu\rho\sigma} = P_{\mu\nu\rho\sigma} + \frac{1}{2}\tilde{K}_{\mu\nu}\tilde{\Omega}_{\rho\sigma} - \frac{1}{4}(\tilde{K}_{\mu\rho}\tilde{\Omega}_{\sigma\nu} + \tilde{K}_{\mu\sigma}\tilde{\Omega}_{\nu\rho} - \tilde{\Omega}_{\mu\rho}\tilde{K}_{\sigma\nu} + \tilde{\Omega}_{\mu\sigma}\tilde{K}_{\nu\rho})$$

che, a differenza di $P_{\mu\nu\rho\sigma}$, gode di <u>tutte</u> le proprietà algebriche di un tensore di curvatura. Avuto riguardo alla (33') e alla (22) si riconosce per $\tilde{R}_{\mu\nu\rho\sigma}$ la seguente forma, ovvia generalizzazione di una espressione classica :

$$(35')\quad \tilde{R}_{\mu\nu\rho\sigma} = \frac{1}{2}\left[(\tilde{\partial}_{\sigma\nu} + \tilde{\partial}_{\nu\sigma})\gamma_{\mu\rho} - (\tilde{\partial}_{\sigma\mu} + \tilde{\partial}_{\mu\sigma})\gamma_{\rho\nu} - (\tilde{\partial}_{\rho\nu} + \tilde{\partial}_{\nu\rho})\gamma_{\mu\sigma} + (\tilde{\partial}_{\rho\mu} + \tilde{\partial}_{\mu\rho})\gamma_{\sigma\nu}\right]$$
$$+ \gamma^{\varepsilon\delta}\left[(\widetilde{\sigma\nu,\varepsilon})(\widetilde{\rho\mu,\delta}) - (\widetilde{\rho\nu,\varepsilon})(\widetilde{\sigma\mu,\delta})\right] .$$

Naturalmente neppure il tensore (35), così come $P_{\mu\nu\rho\sigma}$ (cui si riduce per $\tilde{K}_{\alpha\beta} = 0$ ovvero $\tilde{\Omega}_{\alpha\beta} = 0$) verifica l'identità di <u>Bianchi</u>.

Precisamente, posto che si ha

$$\tilde{\nabla}_{\varepsilon}\tilde{R}_{\mu\nu\rho\sigma} = \tilde{\nabla}_{\varepsilon}\tilde{R}_{\mu\nu\rho\sigma} - \xi_{\varepsilon\mu}\tilde{R}_{0\nu\rho\sigma} - \xi_{\varepsilon\nu}\tilde{R}_{\mu 0\rho\sigma} - \xi_{\varepsilon\rho}\tilde{R}_{\mu\nu 0\sigma} - \xi_{\varepsilon\sigma}\tilde{R}_{\mu\nu\rho 0} ,$$

dalla identità $S\,\nabla_{\varepsilon}\tilde{R}_{\mu\nu\rho\sigma} = 0$, ove S indica la somma dei tre termini ottenuti permutando circolarmente gli indici $\varepsilon, \rho, \sigma$ si trae agevolmente, avuto riguardo alle (32') e $(23)_1$,

$$(36)\quad S\,\tilde{\nabla}_{\varepsilon}(P_{\mu\nu\rho\sigma} + \frac{1}{2}\tilde{K}_{\mu\nu}\tilde{\Omega}_{\rho\sigma}) = -\frac{1}{2}S\,\tilde{\Omega}_{\rho\sigma}(\tilde{\nabla}_{\nu}\tilde{K}_{\mu\varepsilon} + C_{\nu}\tilde{K}_{\mu\varepsilon} - \tilde{\nabla}_{\mu}\tilde{K}_{\nu\varepsilon} - C_{\mu}\tilde{K}_{\nu\varepsilon})$$
$$(\varepsilon, \mu, \nu, \sigma, \rho = 1, 2, 3)$$

nonchè

$$S\tilde{\nabla}_{\varepsilon}\tilde{R}_{\mu\nu\rho\sigma} = -\frac{1}{2} S\tilde{\Omega}_{\rho\sigma}(\tilde{\nabla}_{\nu}\tilde{K}_{\mu\varepsilon} + C_{\nu}\tilde{K}_{\mu\varepsilon} - \tilde{\nabla}_{\mu}\tilde{K}_{\nu\varepsilon} - C_{\mu}\tilde{K}_{\nu\varepsilon}) + \tag{36'}$$
$$\frac{1}{4} S\hat{\nabla}_{\varepsilon}(\tilde{\Omega}_{\mu\rho}\tilde{K}_{\sigma\nu} - \tilde{\Omega}_{\mu\sigma}\tilde{K}_{\nu\rho} + \tilde{K}_{\mu\rho}\tilde{\Omega}_{\nu\sigma} - \tilde{K}_{\mu\sigma}\tilde{\Omega}_{\nu\rho}).$$

Tanto per $\tilde{K}_{\alpha\beta} = 0$ (congruenza rigida), quanto per $\tilde{\Omega}_{\alpha\beta} = 0$ (congruenza normale) si ritrova naturalmente la forma classica $S\tilde{\nabla}_{\varepsilon}P_{\mu\nu\rho\sigma} = S\tilde{\nabla}_{\varepsilon}\hat{R}_{\mu\nu\rho\sigma} = 0$. Nel primo caso perchè la (33) definisce il tensore di curvatura di una V_3 riferita alle coordinate (x^{α}) (varietà quoziente di V_4 rispetto alle linee x^0 = var.) di metrica $\gamma_{\alpha\beta} = \gamma_{\alpha\beta}(x^1, x^2, x^3)$. Nel secondo caso perchè $P_{\mu\nu\rho\sigma} = \tilde{R}_{\mu\nu\rho\sigma}$ è il tensore di curvatura delle varietà normali alla congruenza x^0 = var.

Per quanto riguarda l'inversione di due derivazioni covarianti trasverse è agevole riconoscere direttamente, ovvero per il tramite della (12)[14], che per un qualunque vettore spaziale ($\overset{0}{v} = 0$) si ha [15].

$$\tilde{\nabla}_{\sigma}\tilde{\nabla}_{\rho}v_{\nu} - \tilde{\nabla}_{\rho}\tilde{\nabla}_{\sigma}v_{\nu} = \tilde{\Omega}_{\sigma\rho}\bar{\partial}v_{\nu} - v_{\mu}\tilde{P}^{\mu}{}_{\nu\rho\sigma} . \tag{37}$$

Analogamente per un tensore spaziale.

5. Alcune conseguenze delle (32). Caso di una varietà piatta.

Tenuto conto della posizione (35) e, per la $(32')_2$, della identità $(23)_1$[16] (ove le derivate pfaffiane ∂_{α} possono essere sostituite dalla derivazione covariante trasversa), le (32') si scrivono al modo seguente :

---

(14) Basta tener conto della $(32)_1$ e dell'identità, valida per $\overset{0}{v} = 0$,
$$\nabla_{\sigma}\nabla_{\rho}v_{\nu} = \tilde{\nabla}_{\sigma}\tilde{\nabla}_{\rho}v_{\nu} - \xi_{\sigma\rho}\nabla_{0}v_{\nu} - \xi_{\sigma\nu}\nabla_{\rho}\overset{0}{v} = \tilde{\nabla}_{\sigma}\tilde{\nabla}_{\rho}v_{\nu} - \xi_{\sigma\rho}(\bar{\partial}v_{\nu} - \xi_{\nu}{}^{\varepsilon}v_{\varepsilon}) + \xi_{\sigma\nu}\xi_{\rho}{}^{\varepsilon}v_{\varepsilon}.$$

(15) Cfr. I. Cattaneo Gasparini, Proiezioni dei tensori di curvatura di una varietà riemanniana a metrica iperbolica normale, Rend. di Matem., Vol. XXII, 1-2 (1963), p. 145.

(16) Si notiche la (23) traduce in sostanza le due seguenti proprietà algebriche del tensore di curvatura : $R_{0\nu\rho\sigma} + R_{0\rho\sigma\nu} + R_{0\sigma\nu\rho} = 0$, $R_{0\nu0\sigma} = R_{0\sigma0\nu}$.

$$(38)\quad \begin{cases} \underset{\mu\nu\rho\sigma}{R} = \underset{\mu\nu\rho\sigma}{\tilde{R}} + \frac{1}{H}(\tilde{K}_{\rho\nu}\tilde{K}_{\sigma\mu} - \tilde{K}_{\sigma\nu}\tilde{K}_{\rho\mu}) + \frac{1}{H}(\tilde{\Omega}_{\rho\nu}\tilde{\Omega}_{\sigma\mu} - \tilde{\Omega}_{\sigma\nu}\tilde{\Omega}_{\rho\mu}) + \frac{1}{2}\tilde{\Omega}_{\nu\mu}\tilde{\Omega}_{\rho\sigma} \\ \tilde{\nabla}_\nu \tilde{\Omega}_{\rho\sigma} = \tilde{\nabla}_\rho \tilde{K}_{\sigma\nu} - \tilde{\nabla}_\sigma \tilde{K}_{\rho\nu} - C_\nu \tilde{\Omega}_{\rho\sigma} + C_\rho \tilde{\Omega}_{\sigma\nu} + C_\sigma \tilde{\Omega}_{\nu\rho} - 2 \underset{0\nu\rho\sigma}{R} \\ \tilde{\nabla}_\sigma C_\nu = \frac{1}{2}\bar{\partial}(\tilde{K}_{\sigma\nu} + \tilde{\Omega}_{\sigma\nu}) - \frac{1}{H}(\tilde{K}_{\sigma\rho} + \tilde{\Omega}_{\sigma\rho})(\tilde{K}_\nu{}^\rho + \tilde{\Omega}_\nu{}^\rho) - C_\sigma C_\nu - \underset{0\nu 0\sigma}{R} \end{cases}$$

D'altra parte, applicando l'operazione $\bar{\partial}$ ai due membri della $(32')_2$ si ottiene

$$\tilde{\Omega}_{\rho\sigma}\bar{\partial}C_\nu = \bar{\partial}(\tilde{\nabla}_\rho \zeta_{\sigma\nu} - \tilde{\nabla}_\sigma \zeta_{\rho\nu} - \underset{0\nu\rho\sigma}{R}) - \bar{\partial}\Omega_{\rho\sigma} C_\nu$$

da cui

$$(39)\qquad \tilde{\Omega}^2\bar{\partial}C_\nu = [\bar{\partial}(2\tilde{\nabla}_\rho \zeta_{\sigma\nu} - \underset{0\nu\rho\sigma}{R}) - \bar{\partial}\Omega_{\rho\sigma}C_\nu]\tilde{\Omega}^{\rho\sigma}$$

ove si è posto

$$(40)\qquad \tilde{\Omega}^2 = \tilde{\Omega}_{\rho\sigma}\tilde{\Omega}^{\rho\sigma} .$$

Nella $(38)_1$ gli indici μ , ν , ρ , σ , variando da 1 a 3, non possono essere tutti distinti. Non è pertanto restrittivo supporre $\rho = \mu$ (con che degli ultimi tre termini uno si annulla e due si sommano), in modo che di fatto la $(38)_1$ risulta equivalente a

$$(41)\qquad \underset{\mu\nu\rho\sigma}{R} = \underset{\mu\nu\rho\sigma}{\tilde{R}} + \frac{1}{H}(\tilde{K}_{\rho\nu}\tilde{K}_{\sigma\mu} - \tilde{K}_{\sigma\nu}\tilde{K}_{\rho\mu}) + \frac{3}{H}\tilde{\Omega}_{\nu\mu}\tilde{\Omega}_{\rho\sigma} .$$

In quest'ultima forma essa ben si presta ad esprimere il tensore vortice spaziale mediante il tensore di Riemann di V_H e la metrica spaziale :

$$(42)\quad \begin{cases} 3\tilde{\Omega}^2 = H(\tilde{R}^{\rho\sigma}{}_{\rho\sigma} - R^{\rho\sigma}{}_{\rho\sigma}) + \tilde{K}_\rho{}^\sigma \tilde{K}_\sigma{}^\rho - (\tilde{K}_\sigma{}^\sigma)^2 \\ 3(\tilde{\Omega}_{\rho\sigma})^2 = H(\underset{\rho\sigma\rho\sigma}{\tilde{R}} - \underset{\rho\sigma\rho\sigma}{R}) + (\tilde{K}_{\rho\sigma})^2 - \tilde{K}_{\rho\rho}\tilde{K}_{\sigma\sigma} \end{cases} .$$

La (41) generalizza una relazione analoga ottenuta recentemente da Pirani-Williams nel caso di una congruenza rigida(17). Essa permette di ricavare, in particolare, $\bar{\partial}\tilde{\Omega}_{\rho\sigma}$ mediante $\tilde{\Omega}_{\rho\sigma}$, $\gamma_{\rho\sigma}$ e il tensore di curvatura; sì che in definitiva le $(38)_3$ e (39) vengono complessivamente ad esprimere tutte le derivate prime del vettore di curvatura delle linee x^0 = var. mediante C_ρ stesso, $\tilde{\Omega}_{\rho\sigma}$, $\gamma_{\rho\sigma}$ e il tensore di Riemann.(18)

Se le linee x^0 = var. costituiscono una congruenza rigida $[\dot{\gamma}_{\rho\sigma}=0$ ovvero $\gamma_{\rho\sigma}=\gamma_{\rho\sigma}(x^1,x^2,x^3)]$ le formule precedenti si semplificano notevolmente. In particolare le (39) e $(42)_2$ si scrivono rispettivamente tenuto anche conto della $(22)_2$,

$$(43)\qquad \begin{cases} \tilde{\Omega}^{\nu}\bar{\partial}C_\nu = [\tilde{\nabla}_\rho(\bar{\partial}\tilde{\Omega}_{\sigma\nu})-\bar{\partial}R_{0\nu\rho\sigma}+C_\rho\bar{\partial}\tilde{\Omega}_{\sigma\nu}-C_\nu\bar{\partial}\tilde{\Omega}_{\rho\sigma}]\tilde{\Omega}^{\rho\sigma} \\ 3(\tilde{\Omega}_{\rho\sigma})^2 = 4(\tilde{R}_{\rho\sigma\rho\sigma}-R_{\rho\sigma\rho\sigma}) \end{cases}$$

ove i simboli di Christoffel spaziali, nonchè $\tilde{R}_{\rho\sigma\rho\sigma}$, sono indipendenti dalla variabile x^0: $3\bar{\partial}(\tilde{\Omega}_{\rho\sigma})^2=-4\bar{\partial}R_{\rho\sigma\rho\sigma}$.

Se in più la varietà V_4 è minkowskiana, ovvero $R_{ikmn}=0$ (19), dalle (43) segue nell'ordine, almeno per $\tilde{\Omega}^{\nu}\neq 0$,

$$(44)\qquad \bar{\partial}\tilde{\Omega}_{\rho\sigma}=0 \quad , \quad \bar{\partial}C_\rho=0 \quad ;$$

ciò che basta, per la $(23)_2$, ad assicurare l'esistenza di una funzione delle sole coordinate spaziali x^1, x^2, x^3: $\varphi(x^1,x^2,x^3)$ tale che sia

$$(45)\qquad C_\rho=\frac{\partial\varphi}{\partial x^\rho} \qquad (\rho=1,2,3).$$

(17) Cfr. loc. cit. (4), p. 10.

(18) Per il caso rigido cfr. loc. cit. (4), p. 12.

(19) Meno restrittivamente, per la validità della (44) basta che sia $\bar{\partial}R_{\rho\sigma\rho\sigma}=0$, $\bar{\partial}R_{0\nu\rho\sigma}=0$ $(\nu,\rho,\sigma=1,2,3)$, condizioni largamente soddisfatte nel caso di uno spazio piatto.

D'altra parte perchè una congruenza definisca un gruppo di isometrie ad un parametro è necessario e sufficiente che essa sia rigida e il vettore di curvatura derivi da un potenziale[20].

Si ritrova così il teorema di Herglotz-Noether[21] : in uno spazio piatto ogni congruenza rigida definisce un gruppo di isometrie ad un parametro.

Nelle stesse ipotesi si ha, come dalla (41),

$$3\tilde{\Omega}_{\mu\nu}\tilde{\Omega}_{\rho\sigma} = 4\tilde{R}_{\mu\nu\rho\sigma} \tag{46}$$

mentre le $(38)_{2,3}$ si riducono a

$$\begin{cases} \tilde{\nabla}_\nu \tilde{\Omega}_{\rho\sigma} = - C_\nu \tilde{\Omega}_{\rho\sigma} + C_\rho \tilde{\Omega}_{\sigma\nu} + C_\sigma \tilde{\Omega}_{\nu\rho} \\ \tilde{\nabla}_\sigma C_\nu = -\frac{1}{4} \tilde{\Omega}_{\sigma\rho} \tilde{\Omega}_\nu{}^\rho - C_\sigma C_\rho \end{cases} \tag{47}$$

Le equazioni scritte (47) sono illimitatamente integrabili, come risulta applicando la (37), nè l'identità del Bianchi $S\tilde{\nabla}_\varepsilon \tilde{R}_{\mu\nu\rho\sigma} = 0$ implica, in base alla (46), ulteriori limitazioni per il tensore vortice spaziale.

Si rilevi, per finire, che, mentre classicamente in ogni moto rigido la velocità angolare (relativa) è indipendente dal posto, in relatività ristretta la velocità angolare (assoluta) è, per ogni moto rigido secondo Born, costante solo lungo le linee della congruenza (cioè nel tempo). In base alla $(47)_1$ l'indipendenza dal posto, $\tilde{\nabla}_\nu \tilde{\Omega}_{\rho\sigma} = 0$ si ha solo nei due casi $C_\nu = 0$ (congruenza geodetica), ovvero $\tilde{\Omega}_{\rho\sigma} = 0$ (congruenza normale). In quest'ultimo ca-

(20) Cfr. ad es. Cattaneo, C., Introduzione alla teoria einsteiniana della gravitazione, Vol. II, Roma, Veschi (1961).

(21) Cfr. Herglotz, G., Über den vom Standpunkt des Relativitäts-prinzips aus als starr zu bezeichnenden Körper, Annalen der Physik, t. 31, p. 393 (1910), nonchè Noether, F., Zur Kinematik des starren Körpers in der Relativitätstheorie, Annalen der Physik, t. 31, p. 919(1910).

so è nullo, come dalla (46), il tensore di curvatura spaziale e la congruenza x^0 = var. viene ad ammettere come varietà normali una semplice infinità di iperpiani ecc.

CENTRO INTERNAZIONALE MATEMATICO ESTIVO

(C. I. M. E.)

L. MARIOT

INTERPRETATIONS PHYSIQUES DU QUINZIEME POTENTIEL EN THEORIE PENTADIMENSIONNELLE

INTERPRETATIONS PHYSIQUES DU QUINZIEME POTENTIEL EN THEORIE PENTADIMENSIONNELLE

par L. Mariot

I - La théorie de Jordan-Thiry

1. - La variété riemannienne V_5 :

V_5 est une variété différentiable analogue à V_4 :

1) Dans l'intersection de 2 systèmes de coordonnées admissibles, les coordonnées $x^{\alpha'}$ d'un point x de V_5 dans l'un des systèmes sont des fonctions 2 fois continument différentiables à jacobien non nul des coordonnées x^{α} du même point dans l'autre système.

2) Sur V_5 est définie une métrique riemannienne partout de type hyperbolique normal. Dans un système de coordonnées admissible, on aura :

$$(1) \qquad d\sigma^2 = \gamma_{\alpha\beta}\, dx^{\alpha}\, dx^{\beta} \qquad \alpha, \beta = 0, 1, 2, 3, 4$$

les $\gamma_{\alpha\beta}$ seront de classe C^1 ce sont les potentiels pour le système de coordonnées envisagé.

En chaque point de V_5 on a :

$$d\sigma^2 = -(\omega^0)^2 + (\omega^4)^2 - (\omega^1)^2 - (\omega^2)^2 - (\omega^3)^2$$

où les ω^{α} sont un système de formes de Pfaff linéairement indépendantes.

Le repère $(x, \vec{e}_{\alpha})$ associé est dit orthonormé dans V_5 et l'on a :

$$\vec{dx} = \omega^{\alpha}\vec{e}_{\alpha} \qquad \vec{e}_{\alpha} \cdot \vec{e}_{\beta} = 0 \quad \text{si} \quad \alpha \neq \beta$$

$$(\vec{e}_4)^2 = 1 \qquad (\vec{e}_u)^2 = -1 \qquad (u = 0, 1, 2, 3)$$

V_5 admet un groupe connexe à 1 paramètre d'isométries globales de V_5, à trajectoires z orientées de façon à rendre $d\sigma^2 < 0$, ne laissant invariant aucun point de V_5 (qui se trouve ainsi engendrée par l'ensemble des trajectoires). En outre nous supposons que :

a) les trajectoires sont homéomorphes à un cercle T^1.

b) On peut trouver une variété différentiable V_4 satisfaisant aux mêmes conditions de différentiabilité que V_5, telle qu'il existe un homéomorphisme différentiable de classe C_2 de V_5 sur le produit topologique $V_4 \times T^1$ dans lequel les z s'appliquent sur les cercles facteurs.

V_4 = Variété quotient de V_5 par la relation d'équivalence définie par le groupe d'isométries.

On est conduit à identifier cette variété quotient avec l'espace - temps de la Relativité générale.

Conséquence :

Il existe dans V_5 des coordonnées locales dite adaptées au groupe telles que :

a) Les (x^i) constituent un système de coordonnées locales de V_4. Les variétés $x^o = C^{te}$ sont définies dans V_5 et homéomorphes à V_4.

b) Les $\gamma_{\alpha\beta}$ sont indépendants de x^o le vecteur $\vec{\xi}$ générateur infinitésimal du groupe d'isométries admet pour composantes contravariantes :

$$\xi^i = 0 \qquad \xi^o = 1 \tag{2}$$
$$\vec{\xi}^2 = \gamma_{oo} < 0 \qquad \xi = \sqrt{|\vec{\xi}^2|} \qquad \gamma_{oo} = -\xi^2 .$$

c) Ces coordonnées sont définies modulo les relations :

$$x^{i'} = \psi^{i'}(x^j) \qquad x'^o = x^o + \psi(x^j) \tag{3}$$

Les variétés x^o = Cte d'un système de coordonnées adapté sont les sections de V_5 associées au système. Les sections sont conservées par :

(4) $x^{i'} = \psi^{i'}(x^j)$ $\qquad$ $x'^o = x^o$.

Nous appellerons changement du système des sections (changement de jauge) le changement de coordonnées adaptées :

(5) $x^{i'} = x^i$ $\qquad$ $x'^o = x^o + \psi(x^j)$.

- Soit W_4 une section de V_5. Sur W_4 se trouvent définis des tenseurs qui se transforment selon la loi tensorielle dans le changement (4)

γ_{oo} scalaire

γ_{oi} vecteur covariant

γ_{ij} tenseur symétrique .

- Par contre si un tenseur de W_4 demeure invariant par changement de jauge (5) il est dit intrinsèquement défini sur la variété quotient V_4.

Ex : 1) $\gamma_{oo} = -\xi^2$ scalaire

$$ds^2 = (\omega^4)^2 - (\omega^1)^2 - (\omega^2)^2 - (\omega^3)^2 \quad \text{scalaire}$$

2) $$ds^2 = \left(\gamma_{ij} - \frac{\gamma_{oi}\gamma_{oj}}{\gamma_{oo}}\right) dx^i dx^j$$

$$g_{ij} = \gamma_{ij} - \frac{\gamma_{oi}\gamma_{oj}}{\gamma_{oo}} \qquad g^{ij} = \gamma^{ij}$$

g_{ij} tenseur symétrique

3) si $\gamma_{oi} = \beta \varphi_i \gamma_{oo}$

β constante numérique dont la valeur sera fixée ultérieurement.

Les φ_i dans un changement de jauge se modifient d'un vecteur gradient, donc

$$F_{ij} = \partial_i \varphi_j - \partial_j \varphi_i$$

est intrinsèquement défini sur V_4.

1er stade de l'interprétation :

L'espace quotient V_4 muni de la métrique $ds^2 = g_{ij} dx^i dx^j$ est identifié à l'E - T de la relativité générale.

g_{ij} est le tenseur gravitationnel

φ_i est le potentiel vecteur

F_{ij} le tenseur électromagnétique.

Il restera à interpréter le quinzieme potentiel $\xi^2 = - \gamma_{oo}$.

2. - Les équations de champ dans V_5.

On généralise formellement les équations de la relativité générale et on pose :

$$S_{\alpha\beta} = \textcircled{H}_{\alpha\beta} .$$

Si en un point de V_5, il n'y a ni charge, ni masse $\textcircled{H}_{\alpha\beta} = 0$ [cas unitaire extérieur]. Evidemment il existe alors un champ e. m et un champ gravitationnel.

Si en V_5 existent des charges et des masses, $\textcircled{H}_{\alpha\beta}$ devra décrire au mieux cette répartition. [cas unitaire intérieur].

Nous ne reviendrons pas sur les raisons du choix de $S_{\alpha\beta}$. Nous poserons :

$$S_{\alpha\beta} = R_{\alpha\beta} - \frac{1}{2} \gamma_{\alpha\beta} R \qquad \nabla_\alpha S^{\alpha\beta} = 0 .$$

Les équations du cas unitaire extérieur s'écrivent alors :

(6) $$S_{\alpha\beta} = 0 .$$

S'il existe de la matière pure on écrira :

(7) $$S_{\alpha\beta} = \varrho \, v_\alpha v_\beta .$$

Il est fondamental d'écrire les composantes de $S_{\alpha\beta}$ en fonction des tenseurs définis intrinsèquement sur V_4 pour donner une interprétation physique de la théorie pentadimensionnelle. On obtient :

$$
(8)\left\{
\begin{array}{ll}
\text{a)}\ S_{ij} = \hat{S}_{ij} - \dfrac{\beta^2 \xi^2}{2}\left[\dfrac{1}{4} g_{ij} F_{kl}F^{kl} - F_i{}^k F_{jk}\right] - \dfrac{1}{\xi}\left[\hat{\nabla}_j \partial_i \xi - g_{ij}\hat{\Delta}\xi\right] & \\
 & F^2 = \dfrac{1}{2} F_{kl}F^{kl} \\
\text{b)}\ S_{io} = \dfrac{\beta}{2\xi^2} \hat{\nabla}_j(\xi^3 F^j_i) & \hat{R} = \dfrac{1}{2} g^{ij}\hat{R}_{ij} \\
\text{c)}\ S_{oo} = \dfrac{1}{2}\hat{R} + \dfrac{3}{4}\beta^2 \xi^2 F^2 & \hat{\Delta} = \hat{\nabla}_j(g^{ij}\partial_i)
\end{array}
\right. .
$$

Approfondissons l'interprétation physique déjà amorcée précédemment, en nous astreignant au cas unitaire extérieur :

$$S_{\alpha\beta} = 0$$

entraine :

$$(9) \qquad S_{io} = 0 \qquad \text{donc} \qquad \hat{\nabla}_j(\xi^3 F^j{}_i) = 0$$

l'existence de ξ^3 invite à préciser les grandeurs electromagnétiques du vide :

F_{ij} sera le tenseur Induction Magnétique champ électrique vérifiant

$$\nabla^*_i F^{ij} = 0$$

$H_{ij} = \xi^3 F_{ij}$ sera le tenseur champ Magnétique - induction électrique vérifiant

$$\nabla_i H^{ij} = 0 \quad .$$

Donc $\xi^3 = \varepsilon_o$ joue le rôle de constante diélectrique du vide variable.

Alors le tenseur impulsion-énergie électromagnétique associé au champ est

$$(10) \qquad \tau'_{ij} = \frac{1}{4} g_{ij} H_{kl}F^{kl} - \frac{1}{2}(H_i{}^k F_{jk} + H_{jk}F_i{}^k) = \xi^3 \tau_{ij} \quad .$$

Les équations $S_{ij} = 0$ conduisent à :

(11) $$\hat{S}_{ij} = \frac{\beta^2}{2\xi} \tau'_{ij} + \frac{1}{\xi} (\hat{\nabla}_j \partial_i \xi - g_{ij} \hat{\Delta} \xi) .$$

On est amené à poser $\chi = \frac{\beta^2}{2\xi} = \frac{\chi_o}{\xi}$ <u>facteur de gravitation variable.</u>

Enfin l'équation $S_{oo} = 0$ donne :

(12) $$\frac{1}{2} \hat{R} + \frac{3}{4} \beta^2 \xi^2 F^2 = 0$$

Finalement :

les 4 équations (9) sont les équations de Maxwell .

les 10 équations (11) sont les équations d'Einstein dans V_4 :

$$\hat{S}_{ij} = \chi \left[\tau'_{ij} + \ldots\ldots \right] .$$

Malheureusement le second terme ne reçoit pas ici d'interprétation physique valable.

La quinzième équation de champ (12) détermine les variations de ξ^3 interprété comme constante diélectrique du vide.

II - La métrique conforme

1. - Origine :

Mme Hennequin(2) a montré que les potentiels g_{ij} associés à la métrique induite sur V_4 ne peuvent être assimilés aux potentiels relativistes en 1ère approximation. Elle trouve :

$$(13)\quad \begin{cases} g_{AB} = -\delta_{AB} + \dfrac{2}{c^2}\,\delta_{AB}(\alpha_{oo} - U) + 0(\dfrac{1}{c^4}) & A,\ B = 1,\ 2,\ 3 \\ g_{44} = c^2 - 2U - 2\alpha_{oo} + 0(\dfrac{1}{c^2}) \end{cases}$$

$$\alpha_{oo} = \frac{U}{3} + \frac{Q}{6}$$

U est le potentiel newtonien, Q est un potentiel relatif à une densité fictive $\frac{e^2}{m}$.

Or les potentiels corrects (de Levi-Civita) sont obtenus à la seule condition d'effacer α_{oo} , c'est à dire de poser $\xi = 1$, ce qui revient à la théorie de Kaluza Klein qui consiste à effacer aussi le 15è équation de champ.

La difficulté a été résolue par Mme Hennequin qui a utilisé dans V_4 une métrique conforme en posant :

$$ds^{*2} = \xi\, ds^2 = g^*_{ij} dx^i dx^j$$

$$g^*_{ij} = \xi\, g_{ij} \qquad g^{*ij} = \frac{1}{\xi}\,\gamma^{ij}$$

$$\Gamma^{*k}_{ij} = \Gamma^{k}_{ij} + \frac{1}{2}\left[\delta^k_i \frac{\partial_j \xi}{\xi} + \delta^k_j \frac{\partial_i \xi}{\xi} - g_{ij} g^{kl} \frac{\partial_l \xi}{\xi}\right] .$$

Elle a trouvé :

$$(14)\quad \begin{cases} g^*_{AB} = -\delta_{AB} - \dfrac{2}{c^2}\,\delta_{AB} U + 0(\dfrac{1}{c^4}) \\ g^*_{44} = c^2 - 2U + 0(\dfrac{1}{c^2}) \end{cases} .$$

Les potentiels conformes coincident en 1ère approximation avec les potentiels de Levi-Civita.

2. - Ecriture des équations de champ en métrique conforme.

On admet que F_{ij} sont des données intrinsèques physiques et qu'elles doivent rester invariantes par changement de métrique, donc

$$F^{*}_{ij} = F_{ij} \qquad F^{*ij} = \frac{1}{\xi^2} F^{ij} \quad .$$

On avait

$$\tau'_{ij} = \xi^3 \tau_{ij} = \xi^3 \left(\frac{1}{4} g_{ij} F_{kl} F^{kl} - F_i{}^k F_{jk} \right) \quad .$$

Il viendra :

$$\tau'^{*}_{ij} = \xi^3 \left(\frac{1}{4} g^{*}_{ij} F^{*}_{kl} F^{*kl} - . \right) = \xi^3 \left(\frac{1}{4} \xi \, g_{ij} F_{kl} \frac{1}{\xi^2} F^{kl} \ldots \right)$$

$$\tau'^{*}_{ij} = \frac{1}{\xi} (\xi^3 \tau_{ij}) = \frac{1}{\xi} \tau'_{ij} \qquad \tau'^{*}_{ij} = \xi^2 \tau_{ij} \quad .$$

Un calcul classique donne :

$$\hat{S}_{ij} = \hat{S}^{*}_{ij} + \frac{3}{4} g_{ij} g^{pq} \frac{\partial_p \xi}{\xi} \cdot \frac{\partial_q \xi}{\xi} - \frac{3}{2} \frac{\partial_i \xi}{\xi} \cdot \frac{\partial_j \xi}{\xi} + \frac{1}{\xi} \left[\hat{\nabla}_i \partial_j \xi - g_{ij} \hat{\Delta} \xi \right]$$

Ainsi l'équation 8(a) devient :

$$S_{ij} = \hat{S}^{*}_{ij} + \frac{3}{4} g_{ij} g^{pq} \frac{\partial_p \xi}{\xi} \frac{\partial_q \xi}{\xi} - \frac{3}{2} \frac{\partial_i \xi}{\xi} \frac{\partial_j \xi}{\xi} - \frac{\beta^2}{2} \tau'^{*}_{ij}$$
$$+ \frac{1}{\xi} \left[\hat{\nabla}_i \partial_j \xi - g_{ij} \hat{\Delta} \xi \right] - \frac{1}{\xi} (\hat{\nabla}_i \partial_j \xi - g_{ij} \hat{\Delta} \xi)$$

Or $\quad \frac{\beta^2}{2} = \chi_o \quad .$

Donc on a, dans le cas unitaire extérieur :

15a) $$\hat{S}^{*}_{ij} = \chi_o \tau'^{*}_{ij} - \frac{3}{4} g^{*}_{ij} g^{*pq} \frac{\partial_p \xi}{\xi} \cdot \frac{\partial_q \xi}{\xi} + \frac{3}{2} \frac{\partial_i \xi}{\xi} \frac{\partial_j \xi}{\xi}$$

De nouveaux avantages de la métrique conforme apparaissent :

1) Le coefficient de gravitation χ_o ne dépend plus du 15è potentiel, il est maintenant constant et peut être identifié au coefficient classique.

2) Les dix premières équations de champ (15a) ne font plus intervenir les dérivées secondes du 15 è potentiel.

Pigeaud dans sa thèse (3), a repris les calculs des potentiels en 2è approximation et a montré que seul les potentiels conformes étaient corrects. Il a pu retrouver les équations de mouvement correctes dans une étude très fine en exigeant que la trajectoire pentadimensionnelle d'une particule soit géodésique de la métrique conforme sur V_5. Nous n'aborderons pas cette étude.

Par contre, il reste dans l'équation 15a) les termes

$$- \frac{3}{4} g^*_{ij} g^{*pq} \frac{\partial_p \xi}{\xi} \frac{\partial_q \xi}{\xi} + \frac{3}{2} \frac{\partial_i \xi}{\xi} \frac{\partial_j \xi}{\xi}$$

qui ne reçoivent pas d'interprétation énergétique mais dont la contribution est négligeable en 2è approximation.

III - Le champ mésonique en théorie de Jordan-Thiry

En utilisant le principe variationnel classique il va être possible de donner une interprétation physique nouvelle du 15è potentiel ξ et des termes

$$- \frac{3}{4} g^{*}_{ij} g^{*pq} \frac{\partial_p \xi}{\xi} \cdot \frac{\partial_q \xi}{\xi} + \frac{3}{2} \frac{\partial_i \xi}{\xi} \frac{\partial_j \xi}{\xi} .$$

1. - Extension du principe variationnel au cas intérieur.

Etant donné dans V_4 une chaîne différentiable de dimension 4 , considérons les variations δg_{ij} des potentiels telles que sur la frontière de C , ces variations s'annulent ainsi que celles de leurs dérivées premières. L'intégrale

$$(16) \qquad I = \int_c (\sqrt{-g}\; R_{ij} g^{ij}) dx^1 \wedge dx^2 \wedge dx^3 \wedge dx^4$$

subit alors la variation :

$$(16') \qquad \delta I = \int_c R_{ij}\, \delta q^{ij} dx^1 \wedge dx^2 \wedge dx^3 \wedge dx^4 \qquad q^{ij} = \sqrt{-g}\, g^{ij}$$

$\delta I = 0$ quels que soient les variations δq^{ij} conduisent aux équations de champ $R_{ij} = 0$ du cas extérieur.

On envisage l'extension au cas intérieur de la manière suivante : soit $\mathfrak{M}$ une expression représentant les sources, les équations de champ reçoivent a priori la forme variationnelle :

$$(17) \qquad \delta \int_c (\sqrt{-g}\, g^{ij} R_{ij} - \mathfrak{M}) dx^1 \wedge dx^2 \wedge dx^3 \wedge dx^4 = 0$$

Si $\mathfrak{M}$ ne dépend que des potentiels et non de leurs dérivées premières, (17) s'écrit :

$$\int_c (R_{ij} - \frac{\partial \mathfrak{M}}{\partial q^{ij}})\, \delta q^{ij} dx^1 \wedge dx^2 \wedge dx^3 \wedge dx^4 = 0$$

Soit (18) $R_{ij} = \dfrac{\partial \mathcal{M}}{\partial q^{ij}}$ ou $S_{ij} = \dfrac{1}{\sqrt{-g}} \dfrac{\partial \mathcal{M}}{\partial g^{ij}} = \chi T_{ij}$.

2. - Cas du champ mésonique scalaire - (4)

En relativité restreinte, le méson scalaire $\mathcal{M}^o$ est représenté à l'aide du Lagrangien :

(18) $$\mathcal{L} = g^{ij} \partial_i \varphi \cdot \partial_j \varphi - m^2 \varphi^2 \quad .$$

En relativité générale, nous choisirons :

(19) $$\mathcal{M} = \chi \sqrt{-g}\, \mathcal{L} \quad .$$

Il vient donc :

$$\frac{\partial \mathcal{M}}{\partial g^{ij}} = \chi \sqrt{-g} \frac{\partial \mathcal{L}}{\partial g^{ij}} + \chi \mathcal{L} \frac{\partial \sqrt{-g}}{\partial g^{ij}} \quad .$$

En tenant compte de l'idéntité

$$\frac{\partial \sqrt{-g}}{\partial g^{ij}} = - \frac{1}{2} \sqrt{-g}\; g_{ij}$$

il vient :

$$\frac{\partial \mathcal{M}}{\partial g^{ij}} = \chi \sqrt{-g} \left[\partial_i \varphi \cdot \partial_j \varphi - \frac{1}{2} g_{ij} g^{pq} \partial_p \varphi \cdot \partial_q \varphi + \frac{m^2}{2} \varphi^2 g_{ij} \right]$$

d'où

(20) $$T_{ij} = \partial_i \varphi \cdot \partial_j \varphi - \frac{1}{2} g_{ij} g^{pq} \partial_p \varphi \cdot \partial_q \varphi + \frac{m^2}{2} \varphi^2 g_{ij} \quad .$$

Il est intéressant de noter au passage que $\nabla_i T^i_j = 0$ conduit à $\nabla_i \nabla^i \varphi + \omega^2 \varphi = 0$ l'équation de Klein-Gordon.

3. - Cas de la théorie de Jordan-Thiry.

Il est très instructif de comparer l'équation (15a) et l'équation (20). En effet, en posant :

(21) $$\varphi^2 = \frac{3}{2\chi_o} (\mathrm{Log}\, \xi)^2$$

on a

$$\partial_i \varphi = \sqrt{\frac{3}{2\chi_o}} \frac{\partial_i \xi}{\xi}$$

et T_{ij} donné par (20) s'écrit :

$$T_{ij} = \frac{3}{2\chi_o} \frac{\partial_i \xi}{\xi} \frac{\partial_j \xi}{\xi} - \frac{3}{4\chi_o} g^*_{ij} g^{*pq} \frac{\partial_p \xi}{\xi} \frac{\partial_q \xi}{\xi} + \text{termes de masses}$$

(22) $$\chi_o T_{ij} = \frac{3}{2} \frac{\partial_i \xi}{\xi} \frac{\partial_j \xi}{\xi} - \frac{3}{4} g^*_{ij} g^{*pq} \frac{\partial_p \xi}{\xi} \frac{\partial_q \xi}{\xi} + \text{termes de masses}$$

A ce moment l'équation (15a) s'écrit en l'absence de termes de masse :

(23) $$\hat{S}^*_{ij} = \chi_o \left[\tau'_{ij} + T'_{ij} \right]$$

$$T'_{ij} = \frac{3}{2} \frac{\partial_i \xi}{\xi} \frac{\partial_j \xi}{\xi} - \frac{3}{4} g^*_{ij} g^{*pq} \frac{\partial_p \xi}{\xi} . \frac{\partial_q \xi}{\xi} \quad .$$

Ainsi le 15è potentiel ξ s'interprète ici comme le champ mésonique et la 15è équation de champ

$$\Delta (\log \xi) + \chi H^*_{ij} F^{*ij} = 0$$

est à comparer à une équation de Klein-Gordon.

Evidemment les termes de masse se mettent dans les seconds membres des équations de champ.

(1) A. Lichnerowicz - Théories de la gravitation et de l'électromagnétisme - Masson, Paris.

(2) F. Hennequin - Etude Mathématique des approximations en relativité générale et en théorie unitaire de Jordan-Thiry, Bull. Scient. Commiss. Trav. hist. et scient. t. I. 1956 2è partie, p. 73 - 154 - Paris, Gauthier-Villars.

(3) P. Pigeaud - Généralisation des schèmas matière pure et fluide parfait en théorie pentadimensionnelle de Jordan-Thiry. Annales de l'Institut Fourier - Grenoble, 1963.

(4) L. Mariot - P. Pigeaud - CR - Ac. Sc. 1963 t. 257 p. 621-623.

CENTRO INTERNAZIONALE MATEMATICO ESTIVO

(C. I. M. E.)

G. CARICATO

SUL PROBLEMA DI CAUCHY PER LE EQUAZIONI GRAVITAZIONALI NEL VUOTO

SUL PROBLEMA DI CAUCHY PER LE EQUAZIONI GRAVITAZIONALI NEL VUOTO

di

G. CARICATO

Nello studio del problema di Cauchy per le equazioni gravitazionali nel vuoto, un primo teorema di unicità, con semplici ipotesi di differenziabilità per i dati di Cauchy, è dovuto a K. Stellmacher [1].

Successivamente A. Lichnerowicz [2] mostrò che il problema si spezza, in modo naturale, in un problema concernente i dati iniziali (che devono soddisfare a certe condizioni di compatibilità) e nel problema di evoluzione vero e proprio; e stabilì teoremi di esistenza e unicità nell'ipotesi che i dati di Cauchy fossero analitici. Y. Bruhat confermò i suddetti teoremi [3] supponendo i dati di Cauchy semplicemente differenziabili, e dette forma invariantiva alle condizioni di compatibilità [4].

Benchè perfettamente esaurienti da un punto di vista matematico, le trattazioni suddette, nel complesso meccanismo delle equazioni gravitazionali e della indeterminazione che la libera scelta delle coordinate induce sulle derivate seconde dei potenziali gravitazionali, lasciano un poco in ombra il significato intrinseco del problema e delle successive fasi della sua risoluzione.

---

(1) K. Stellmacher, Zum Anfangswertproblem der Gravitationsgleichungen, Mathematische Annalen , 115, Berlin 1938
K. Stellmacher, Ausbreitungsgesetze für charakteristiche Singularitaten der Gravitationsgleichungen, Mathematische Annalen 115 , Berlin 1938

(2) A. Lichnerowicz, Problèmes globaux en mècanique relativiste. Actualités Scientifiques Industrielles, Paris 1939.

(3) Y. Fourés-Bruhat, Théorème d'existence pour certains systèmes d'équations aux dérivées partielles non linéaires, Acta Mathematica 88, 1952

(4) Y. Fourés-Bruhat, Sur l'integration des equations de la Relativité Générale, Journal of rational mechanics and analysis, vol. 5, 1956.

Mediante l'uso sistematico della tecnica delle proiezioni [5] ho cercato di chiarire tale significato intrinseco ritrovando, concordemente a M.me Bruhat, quali siano gli oggetti geometrici che costituiscono i dati di Cauchy, e precisando inoltre quali siano gli oggetti geometrici che le equazioni di evoluzione valgono a determinare [6].

1. Equazioni gravitazionali in forma relativa a un determinato riferimento fisico.

Assegnato in uno spaziotempo einsteiniano V_4 un sistema di coordinate locali x^i fisicamente ammissibili, e quindi un determinato riferimento fisico S, si considerino le proiezioni naturali del tensore di curvatura contratto R_{jm} [7]

---

[5] C. Cattaneo, Proiezioni naturali e derivazione trasversa in una varietá riemanniana a metrica iperbolica normale, Annali di Matematica pura ed applicata (IV) Vol. XLVIII, 1959.

[6] G. Caricato, Sul problema di Cauchy per le equazioni gravitazionali nel vuoto, Rendiconti di Matematica (3-4) Vol. 22, 1963.

[7] Ida Cattaneo Gasparini, Projections naturelles des tenseurs de courbure d'une varieté V_{n+1} à métrique hyperbolique normale, Comptes rendus Académie des Sciences, t. 252, 1961.

$$(1)\quad \begin{cases} \mathcal{P}_{\Sigma\Sigma}(R_{jm}) = \tilde{R}^*_{jm} + \frac{1}{4}\tilde{K}^i_i(\tilde{K}_{mj}+\tilde{\Omega}_{mj}) - \frac{1}{2}(\tilde{K}^{\ i}_m+\tilde{\Omega}^{\ i}_m)\tilde{K}_{ij} + \frac{1}{2}\gamma^4\partial_4(\tilde{K}_{mj}+\tilde{\Omega}_{mj}) + \\ \qquad + \frac{1}{2}\tilde{\Omega}^i_j\tilde{\Omega}_{im} - C_j C_m - \tilde{\nabla}^*_m C_j \\ \mathcal{P}_{\Sigma\Theta}(R_{jm}) = \gamma_m\left[\frac{1}{2}\left\{\tilde{\nabla}^*_j\tilde{K}^i_i - \tilde{\nabla}^*_h(\tilde{K}^{\ h}_j+\tilde{\Omega}^{\ h}_j)\right\} + C_i\tilde{\Omega}^{\ i}_j\right] \\ \mathcal{P}_{\Theta\Sigma}(R_{jm}) = \gamma_j\left[\frac{1}{2}\left\{\tilde{\nabla}^*_m\tilde{K}^i_i - \tilde{\nabla}^*_i(\tilde{K}^{\ i}_m+\tilde{\Omega}^{\ i}_m)\right\} + C^i\tilde{\Omega}_{im}\right] \\ \mathcal{P}_{\Theta\Theta}(R_{jm}) = \gamma_j\gamma_m\left[-\frac{1}{2}\gamma^4\partial_4\tilde{K}^i_i - \frac{1}{4}\tilde{K}^{ij}\tilde{K}_{ij} + \frac{1}{4}\tilde{\Omega}^{ij}\tilde{\Omega}_{ij} + C^iC_i + \tilde{\nabla}^*_i C^i\right]. \end{cases}$$

In virtú delle (1) lo scalare di curvatura

$$R = g^{jm}(\mathcal{P}_{\Sigma\Sigma} + \mathcal{P}_{\Sigma\Theta} + \mathcal{P}_{\Theta\Sigma} + \mathcal{P}_{\Theta\Theta})R_{jm}$$

assume la forma [8]

$$(2)\quad R = \tilde{R}^* + \frac{1}{4}\left[(\tilde{K}^i_i)^2 + \tilde{K}^{ij}\tilde{K}_{ij} + \tilde{\Omega}^{ij}\tilde{\Omega}_{ij} + \gamma^4\partial_4(\tilde{K}^i_i) - 2(\tilde{\nabla}^*_i C^i + C^iC_i)\right.$$

con

$$(3)\quad \begin{cases} \tilde{R}^* = g^{jm}\tilde{R}^*_{jm} = g^{jm}(\tilde{P}^*_{jm} - \frac{1}{2}\tilde{K}^r_j\tilde{\Omega}_{rm}) = \gamma^{\alpha\beta}\tilde{P}^*_{\alpha\beta} \\ \qquad (\tilde{P}^*_{4r} = 0 = \tilde{K}_{4r} = \tilde{\Omega}_{4r}) \\ \tilde{P}^*_{\alpha\beta} = -\tilde{\partial}_\beta\widetilde{\left\{{\varsigma \atop \alpha\varsigma}\right\}} + \tilde{\partial}_\varsigma\widetilde{\left\{{\varsigma \atop \beta\alpha}\right\}} - \widetilde{\left\{{\sigma \atop \alpha\varsigma}\right\}}\cdot\widetilde{\left\{{\varsigma \atop \beta\sigma}\right\}} + \widetilde{\left\{{\sigma \atop \beta\alpha}\right\}}\cdot\widetilde{\left\{{\varsigma \atop \varsigma\sigma}\right\}}. \end{cases}$$

Ciò posto, si prendano in esame le equazioni gravitazionali

$$(4)\quad G_{jm} \equiv R_{jm} - \frac{1}{2}Rg_{jm} = 0,$$

e se ne eseguano le proiezioni naturali.

---

(8) La convenzione che si adotterà per gli indici è la seguente: le lettere greche rappresenteranno gli indici variabili da 1 a 3, quelle latine gli indici variabili da 1 a 4.

Se si pone

$$(5)\quad \begin{cases} s_{jm} = \mathcal{P}_{\Sigma\Sigma}(R_{jm}) \quad (s_{4j} = 0)\,, \quad \mathcal{P}_{\Sigma\theta}(Rjm) = S_i\, \gamma_m \\ S_\alpha = \frac{1}{2}\left[\widetilde{\nabla}^*_\alpha(\widetilde{K}^i_i) - \widetilde{\nabla}^*_h(\widetilde{K}^h_\alpha + \widetilde{\Omega}^h_\alpha)\right] + C^\beta\, \widetilde{\Omega}_{\beta\alpha} \quad (S_4 = 0) \\ I = \frac{1}{4}\left[(\widetilde{K}^\alpha_\alpha)^2 - \widetilde{K}^{\alpha\beta}\, \widetilde{K}_{\alpha\beta} + 3\, \widetilde{\Omega}^{\alpha\beta}\, \widetilde{\Omega}_{\alpha\beta}\right] \end{cases},$$

le (4) equivalgono al seguente sistema di tre equazioni tensoriali:

$$(6)\quad \begin{cases} s_{\alpha\varsigma} - \frac{1}{2} R\, \gamma_{\alpha\varsigma} = 0 \\ S_\alpha = 0 \\ \widetilde{R}^* + I = 0\,. \end{cases}$$

In modo analogo, poichè si ha

$$(7)\quad G^j_m = (\mathcal{P}_{\Sigma\Sigma} + \mathcal{P}_{\Sigma\theta} + \mathcal{P}_{\theta\Sigma} + \mathcal{P}_{\theta\theta})\, G^j_m \equiv$$

$$s^j_m - \frac{1}{2} R\, \gamma^j_m + S^j \gamma_m + \gamma^j S_m + \frac{1}{2}(\widetilde{R}^* + I)\, \gamma^j \gamma_m\,,$$

le identità differenziali cui soddisfa il tensore gravitazionale G_{ik},

$$(8)\quad \nabla_j\, G^j_m \equiv 0\,,$$

possono tradursi nelle seguenti:

$$
(8)' \begin{cases} \mathcal{P}_{\Sigma}(\nabla_j G^j_m) \equiv \tilde{\nabla}^*_j s^j_m + C_h s^h_m - \frac{1}{2} R C_m - \frac{1}{2} \tilde{\nabla}^*_m R + \frac{1}{2} \tilde{K}^\alpha_\alpha S_m + \gamma^4 \partial_4 S_m + \\ \qquad + \tilde{\Omega}_{hm} S^h + \frac{1}{2}(\tilde{R}^* + I) C_m \equiv 0 , \\ \mathcal{P}_{\Theta}(\nabla_j G^j_m) \equiv \frac{1}{2}(\tilde{K}_{jt} + \tilde{\Omega}_{jt}) s^{jt} - \frac{1}{4} R \tilde{K}^\alpha_\alpha + \tilde{\nabla}^*_j S^j + 2 C_j S^j + \frac{1}{4} \tilde{K}^\alpha_\alpha (\tilde{R}^* + I) + \\ \qquad + \frac{1}{2} \gamma^4 \partial_4 (\tilde{R}^* + I) \equiv 0 . \end{cases}
$$

2. Altra forma delle equazioni gravitazionali nel vuoto

Com'è noto, le (4) sono equivalenti alle equazioni

$$
(9) \qquad R_{jm} = 0 .
$$

Si può verificare però facilmente che le (9) equivalgono anche al sistema di equazioni tensoriali

$$
(10) \begin{cases} s_{\alpha\beta} \equiv \gamma_{\alpha\varsigma} R^\varsigma_\beta + \gamma_\beta \gamma^4 \gamma_{\alpha\varsigma} R^\varsigma_4 \equiv \tilde{R}^*_{\alpha\beta} + \frac{1}{4} K^i_i (\tilde{K}_{\beta\alpha} + \tilde{\Omega}_{\beta\alpha}) - \frac{1}{2} (\tilde{K}^i_\beta + \tilde{\Omega}^i_\beta) \tilde{K}_{i\alpha} + \\ + \frac{1}{2} \gamma^4 \partial_4 \tilde{K}_{\alpha\beta} - \tilde{\nabla}^*_\beta C_\alpha + \frac{1}{2} \gamma^4 \partial_4 \tilde{\Omega}_{\beta\alpha} + \frac{1}{2} \tilde{\Omega}^i{}_\alpha \tilde{\Omega}_{i\beta} - C_\alpha C_\beta = 0 , \\ S_\beta \equiv - \gamma^4 \gamma_{\beta\varsigma} R^\varsigma_4 \equiv - \gamma^4 (R_{\beta 4} + \gamma_\beta \gamma^4 R_{44}) \equiv \frac{1}{2} \left[\tilde{\nabla}^*_\beta \tilde{K}^i_i - \tilde{\nabla}^*_\mu (\tilde{K}_\beta{}^\mu + \tilde{\Omega}_\beta{}^\mu) \right] + C^\alpha \tilde{\Omega}_{\alpha\beta} = 0 , \\ \tilde{R}^* + I \equiv 2 \gamma^4 \gamma_r G^r_4 \equiv \tilde{R}^* + \frac{1}{4} \left[(\tilde{K}^i_i)^2 - \tilde{K}^{\alpha\beta} \tilde{K}_{\alpha\beta} + 3 \tilde{\Omega}^{\alpha\beta} \tilde{\Omega}_{\alpha\beta} \right] = 0 . \end{cases}
$$

3. Formulazione intrinseca del problema di evoluzione.

Il problema di evoluzione per le equazioni gravitazionali nelle regioni vuote dello spaziotempo può avere la seguente formulazione intrinseca:

a) In una varietà differenziabile V_4 sia assegnata una porzione di ipersuperficie regolare Σ .

b) Si scelga un arbitrario campo ausiliario di vettori controvarianti $\gamma^i(x)$ definiti in tutto un intorno 4-dimensionale di Σ , dotati di derivate parziali fino al 4^0 ordine, continue, limitate e lipschitziane, con l'unica condizione che, nei punti di $\bar{\Sigma}$, essi non siano tangenti alla $\bar{\Sigma}$ medesima.

c) Si introduca anche un campo ausiliario di vettori covarianti $\eta_i(x)$ dotati di derivate parziali fino al 4^0 ordine, continue, limitate e lipschitziane, legati ovunque al precedente campo $\gamma^i(x)$ dalla condizione

$$\eta_i \cdot \gamma^i = -1 \tag{11}$$

e tali da soddisfare su $\bar{\Sigma}$ alla relazione

$$\eta_i \bar{\xi}^i = 0 , \tag{12}$$

$\bar{\xi}^i$ essendo un generico vettore controvariante tangente a $\bar{\Sigma}$.

(Ciò significa che nei punti di $\bar{\Sigma}$ le giaciture individuate dai vettori η_i risultano tangenti alla $\bar{\Sigma}$).

d) Si adotti un sistema di coordinate "adattate" al campo $\underline{\gamma}(x)$ e alla ipersuperficie iniziale $\bar{\Sigma}$, cioè tali che linee del campo γ^i abbiano equazioni x^α = cost., x^4 = var., e l'ipersuperficie $\bar{\Sigma}$ equazione $x^4 = 0$. Con tale scelta risulta $\gamma^\alpha = 0$ ovunque e $\bar{\xi}^4 = 0$ su $\bar{\Sigma}$.

e) Si suppongano assegnati, nei punti di $\bar{\Sigma}$, due tensori doppi simmetrici $\bar{\gamma}_{ik}$, $\bar{\varphi}_{ik}$ dotati di derivate parziali fino agli ordini rispettivamente 5^0 e 4^0, continue, limitate e lipschitziane, verificanti ivi, in coordinate adattate, le condizioni

$$\begin{cases} \bar{\gamma}_{i4} = 0 , \qquad \bar{\gamma}_{\alpha\beta}\bar{\xi}^\alpha\bar{\xi}^\beta > 0 \\ \qquad\qquad\qquad\qquad (\text{su } \bar{\Sigma}) \\ \bar{\varphi}_{i4} = 0 \end{cases} \tag{13}$$

nonchè le relazioni differenziali [(9)]

$$
(14)\qquad \begin{cases} \tilde{\nabla}^*_\alpha(\bar{\varphi}^\mu_\mu) - \tilde{\nabla}^*_\mu(\bar{\varphi}^\mu_\alpha) = 0 \\ \tilde{R}^* + \frac{1}{4}\left[(\bar{\varphi}^\mu_\mu)^2 - \bar{\varphi}^{\alpha\beta}\bar{\varphi}_{\alpha\beta}\right] = 0 . \end{cases}
$$

f) Con tali dati si determini, in un intorno 4-dimensionale W di Σ , un campo di tensori doppi simmetrici γ_{ij} ivi soggetti alla condizione $\gamma_{i4}=0$, tali da soddisfare in W il sistema differenziale [(10)] (equazioni di evoluzione)

$$
\begin{aligned}
2s_{\alpha\beta} \equiv (\gamma^4)^2 \Big[& \partial_4\partial_4\gamma_{\alpha\beta} - \eta_\alpha\eta_\beta\gamma^{s\sigma}\partial_4\partial_4\gamma_{s\sigma} + \gamma^{s\sigma}\eta_s(\eta_\alpha\partial_4\partial_4\gamma_{\beta\sigma} + \\
& + \eta_\beta\partial_4\partial_4\gamma_{\alpha\sigma} - \eta_\sigma\partial_4\partial_4\gamma_{\alpha\beta})\Big] + \\
(15)\qquad - \gamma^{s\sigma}\Big[& \partial_s\partial_\sigma\gamma_{\alpha\beta} + \partial_\alpha\partial_\beta\gamma_{s\sigma} - \partial_s\partial_\beta\gamma_{\alpha\sigma} - \partial_s\partial_\alpha\gamma_{\beta\sigma} + \eta_s\gamma^4(\partial_\sigma\partial_4\gamma_{\alpha\beta} + \\
& - \partial_\beta\partial_4\gamma_{\alpha\sigma} - \partial_\alpha\partial_4\gamma_{\beta\sigma})\Big] + \\
- \gamma^{s\sigma}\Big[& \eta_\alpha\gamma^4(\partial_\beta\partial_4\gamma_{s\sigma} - \partial_s\partial_4\gamma_{\beta\sigma}) + \eta_\beta\gamma^4(\partial_\alpha\partial_4\gamma_{s\sigma} - \partial_s\partial_4\gamma_{\alpha\sigma}) \Big] - h_{\alpha\beta} = 0
\end{aligned}
$$

---

(9) Ovunque si ripresentino enti e formule già incontrate nelle pagine precedenti, nei quali interveniva il campo di vettori covarianti $\gamma_i(x)$, si intende eseguita, d'ora in poi, la sostituzione di tale campo col campo di vettori covarianti $\eta_i(x)$ introdotto poc'anzi.

(10) Nel sistema (15) le $h_{\alpha\beta}$ rappresentano funzioni razionali ben determinate ma qui non esplicitate, di η_i, γ^i, $\partial_r\eta_i$, $\partial_s\gamma^i$, $\partial_s\partial_r\eta_i$, γ_{ij}, $\partial_r\gamma_{ij}$.
Le funzioni $\gamma^{s\sigma}$ sono definite dalle equazioni

$$\gamma^{s\sigma}\gamma_{\sigma\tau} = \delta^s_\tau \quad .$$

e su $\bar{\Sigma}$ le condizioni iniziali

$$(16)\quad \begin{cases} \gamma_{ij} = \bar{\gamma}_{ij} \\ \gamma^4 \partial_4 \gamma_{ij} = \bar{\psi}_{ij} \end{cases} \quad \text{-----------------} \bar{\Sigma}$$

4. Analisi delle equazioni di evoluzione

Si mostrerà ora che ogni soluzione del problema di evoluzione soddisfa pure, nel medesimo intorno W di $\bar{\Sigma}$, le equazioni che si ottengono dalle $(10)_2$, $(10)_3$ sostituendo in esse al campo di vettori covarianti $\gamma_i(x)$ il campo $\eta_i(x)$

$$(17)\quad \begin{matrix} S_\alpha = 0 \\ \widetilde{R}^* + I = 0 \end{matrix} \quad \text{----------------} \; W$$

Infatti, ammessa l'esistenza di una soluzione delle equazioni di evoluzione (15), che su $\bar{\Sigma}$ soddisfi le (16), si ha per una tale soluzione,

$$(18)\quad \gamma^{\alpha\beta} s_{\alpha\beta} \equiv s^\alpha_\alpha \equiv \widetilde{R}^* + \frac{1}{4}(\widetilde{K}^\alpha_\alpha)^2 - \frac{1}{2}\widetilde{K}^{\alpha\beta}\widetilde{K}_{\alpha\beta} + \frac{1}{2}\gamma^{\alpha\beta}\gamma^4\partial_4(\widetilde{K}_{\beta\alpha} + \widetilde{\Omega}_{\beta\alpha}) - C^\alpha C_\alpha + - \widetilde{\nabla}^*_\alpha C^\alpha + \frac{1}{2}\widetilde{\Omega}^{\alpha\beta}\widetilde{\Omega}_{\alpha\beta} = 0.$$

Dalla (18), in virtù delle (2), $(5)_4$ e dell'identità facilmente verificabile

$$(19)\quad \widetilde{K}^{\alpha\beta} = -\gamma^4\partial_4\gamma^{\alpha\beta},$$

si trae

$$(20)\quad R = -(\widetilde{R}^* + I).$$

Si prendano inoltre in esame le identità $(8)'$. Sostituendo in esse, al generico tensore γ_{ik} la soluzione poc'anzi scelta e al campo di vettori covarianti $\gamma_i(x)$ il campo $\eta_i(x)$, si ha

$$(21)\quad \begin{cases} \gamma^4\partial_4 S_\alpha = -\frac{1}{2}\ \widetilde{\nabla}^*_\alpha(\widetilde{R}^* + I) - \frac{1}{2}\widetilde{K}^\beta{}_\beta S_\alpha - \widetilde{\Omega}_{h\alpha} S^h - (\widetilde{R}^* + I)C_\alpha \\ \frac{1}{2}\gamma^4\partial_4(\widetilde{R}^* + I) = -\ \widetilde{\nabla}^*_\alpha S_\alpha - 2C_\alpha S^\alpha - \frac{1}{2}\widetilde{K}^\alpha{}_\alpha(\widetilde{R}^* + I)\,. \end{cases}$$

Le (21), interpretate come equazioni alle derivate parziali nelle funzioni incognite S_α, $\widetilde{R}^* + I$, possono assumere la seguente forma normale rispetto alla variabile x^4:

$$(21)'\quad \begin{cases} \gamma^4\partial_4\, S_\alpha = -\,\eta_\alpha D + H_\alpha \\ \gamma^4\partial_4(\widetilde{R}^* + I) \ = 2D \end{cases}$$

con

$$H_\alpha(x) \equiv -\frac{1}{2}\partial_\alpha(\widetilde{R}^* + I) - \frac{1}{2}\widetilde{K}^\beta{}_\beta\, S_\alpha + \widetilde{\Omega}_{\alpha h}S^h - (\widetilde{R}^* + I)C_\alpha\,,$$

$$N(x) = -\ \gamma^{\alpha\beta}\partial_\alpha S_\beta + \gamma^{\alpha\beta}\left\{{\varsigma \atop \alpha\beta}\right\}^* S_\varsigma - 2C_\alpha S^\alpha - \frac{1}{2}\widetilde{K}_\alpha^{\ \alpha}\,(\widetilde{R}^* + I);$$

$$D(x) = \frac{-\ \gamma^{\alpha\beta}\eta_\alpha H_\beta + N(x)}{1 - \gamma^{\alpha\beta}\eta_\alpha\eta_\beta}$$

Le (21)' ammettono ovviamente, nell'intorno W di $\bar{\bar{\Sigma}}$, la soluzione nulla

(22) $\qquad S_\alpha = 0 \quad , \quad \widetilde{R}^* + I = 0$ ---------------------W

e d'altra parte questa è l'unica che si annulli sull'ipersuperficie $\bar{\bar{\Sigma}}$ se i coefficienti delle (21)' sono dotati di derivate prime continue.[11].

[11] J. Schauder Lwòw, Cauchy'sches Problem für partielle differentialgheichungen erster ordnung. Anwendung einiger sich auf die Absolutbeträge der Lösungen bezichenden Abschätzungen, Commentarii Mathematici Helvetici, vol. 9 MCMXXXVI/VII.

Pertanto ogni soluzione del problema di evoluzione rappresenta una soluzione dell'intero sistema (10), nel quale al campo γ_i sia stato sostituito il campo η_i.

Se si ricorda che il campo ausiliario di vettori covarianti η_i soddisfa su $\bar{\Sigma}$ la condizione (2), ossia, in coordinate adattate,

$$(\eta_\alpha)_{x^4=0} = 0$$

si riconosce che le equazioni (17), valutate su $\bar{\Sigma}$, forniscono due relazioni differenziali invariantive tra i dati tensoriali di Cauchy, e si identificano con le condizioni di compatibilità (14).

Esplicando la dipendenza degli operatori che in esse compaiono dagli stessi dati di Cauchy, le (14) possono scriversi nel seguente modo:

$$\bar{\gamma}^{\mu\nu}(\partial_\alpha \bar{\varphi}_{\mu\nu} - \partial_\mu \bar{\varphi}_{\nu\alpha}) = (\frac{1}{2}\partial_\nu \log\bar{\gamma} - \bar{\gamma}^{\mu\beta}\partial_\beta \bar{\gamma}_{\mu\nu})\bar{\varphi}^\nu{}_\alpha + \frac{1}{2}\partial_\alpha \bar{\gamma}_{\mu\nu}\cdot\bar{\varphi}^{\mu\nu}$$

$$(14)' \quad \bar{\gamma}^{\alpha\beta}\bar{\gamma}^{\varsigma\tau}(\partial_\alpha\partial_\beta\bar{\gamma}_{\varsigma\tau} - \partial_\alpha\partial_\varsigma\bar{\gamma}_{\beta\tau}) + \bar{\gamma}^{\alpha\beta}\bar{\psi}_{\alpha\beta} + \frac{1}{2}\partial_\nu \log\bar{\gamma}\,(\frac{1}{2}\bar{\gamma}^{\nu\sigma}\partial_\sigma \log\bar{\gamma} + \partial_\varsigma \bar{\gamma}^{\varsigma\nu}) =$$

$$= \frac{1}{4}\left[(\bar{\varphi}^\alpha{}_\alpha)^2 - \bar{\varphi}^{\alpha\beta}\bar{\varphi}_{\alpha\beta}\right],$$

con $\bar{\psi}_{\alpha\beta}$ forme quadratiche in $\partial_\mu \bar{\gamma}_{\varsigma\sigma}$.

Il sistema (15), poichè risulta per ipotesi $\eta_4 \neq 0$, può assumere la seguente forma normale rispetto alla variabile x^4:

$$(15)' \quad \partial_4\partial_4 \gamma_{\varsigma\sigma} = \mathcal{L}_{\varsigma\sigma} + Q_{\varsigma\sigma}.$$

Le funzioni $\mathcal{L}_{\varsigma\sigma}$ sono polinomi lineari in $\partial_\mu\partial_r\gamma_{\mu\nu}$ e $Q_{\varsigma\sigma}$ sono polinomi quadratici in $\partial_i\gamma_{\mu\nu}$.

Limitandosi a supporre analitici tutti i dati del problema, il sistema (15)', in virtù del teorema di Cauchy-Kowalesky ammette un'unica soluzione γ_{ik} (γ_{4i} = 0) che soddisfa su $\bar{\Sigma}$ le condizioni (16).

E ponendo $g_{ik} = \gamma_{ik} - \eta_i \eta_k$, viene ad essere individuato univocamente il tensore metrico g_{ik} della varietà V_4.

È immediato poi costatare che il campo di vettori covarianti η_i fornisce la rappresentazione covariante del campo vettoriale $\gamma^i(x)$.

La risoluzione del problema di evoluzione permette dunque di individuare simultaneamente la metrica di V_4 e, nello spaziotempo così costituito, un ben determinato riferimento S .

Le linee orarie delle particelle che individuano il riferimento S sono fornite dalla congruenza di linee del genere tempo individuata dal campo $\underline{\gamma}$.

Si osservi che, disponendo dell'arbitrarietà con la quale è stato introdotto in V_4 il campo di vettori covarianti η_i, nulla impedisce di scegliere le prime tre componenti η_α identicamente nulle. Fatta tale scelta si verificano le seguenti circostanze:

I. La soluzione $\gamma_{\alpha\beta}$ del problema di evoluzione fornisce la metrica dell'ipersuperficie Σ di equazione x^4 = cost.

II. Le condizioni di compatibilità (14) assegnano la legge con la quale la metrica anzidetta inizia ad evolversi nel tempo.